An AGI Brain for a Robot

An AGI Brain for a Robot

John H Andreae

with Robot Cats by Gillian M Andreae

Academic Press is an imprint of Elsevier
125 London Wall, London EC2Y 5AS, United Kingdom
525 B Street, Suite 1650, San Diego, CA 92101, United States
50 Hampshire Street, 5th Floor, Cambridge, MA 02139, United States
The Boulevard, Langford Lane, Kidlington, Oxford OX5 1GB, United Kingdom

© 2021 Elsevier Inc. All rights reserved.

No part of this publication may be reproduced or transmitted in any form or by any means, electronic or mechanical, including photocopying, recording, or any information storage and retrieval system, without permission in writing from the publisher. Details on how to seek permission, further information about the Publisher's permissions policies and our arrangements with organizations such as the Copyright Clearance Center and the Copyright Licensing Agency, can be found at our website: www.elsevier.com/permissions.

This book and the individual contributions contained in it are protected under copyright by the Publisher (other than as may be noted herein).

Notices
Knowledge and best practice in this field are constantly changing. As new research and experience broaden our understanding, changes in research methods, professional practices, or medical treatment may become necessary.

Practitioners and researchers must always rely on their own experience and knowledge in evaluating and using any information, methods, compounds, or experiments described herein. In using such information or methods they should be mindful of their own safety and the safety of others, including parties for whom they have a professional responsibility.

To the fullest extent of the law, neither the Publisher nor the authors, contributors, or editors, assume any liability for any injury and/or damage to persons or property as a matter of products liability, negligence or otherwise, or from any use or operation of any methods, products, instructions, or ideas contained in the material herein.

Library of Congress Cataloging-in-Publication Data
A catalog record for this book is available from the Library of Congress

British Library Cataloguing-in-Publication Data
A catalogue record for this book is available from the British Library

ISBN 978-0-323-85254-8

For information on all Academic Press publications
visit our website at https://www.elsevier.com/books-and-journals

Publisher: Nikki Levy
Acquisitions Editor: Natalie Farra
Editorial Project Manager: Sam Young
Production Project Manager: Punithavathy Govindaradjane
Cover Designer: Richard Roberts
Robot cat illustrations: Gillian Andreae ©

Typeset by SPi Global, India

Contents

This book provides a Java computer program and output data files via a companion website: https://www.elsevier.com/books-and-journals/book-companion/9780323852548.

List of Figures

Preface

The writing of this book was triggered by a remark of Dan Dennett. I was much encouraged by his approval 2 years later. Generous support for publication of the book was also given by Michael Arbib and Ian Witten. David Hill, Andy Barto, my son Peter, and an anonymous reviewer sharpened my ideas. My daughter Gillian did a lot more than provide the cats. She and Suzanne Brown showed me how to cater for the general reader. My wife, Molly, spent many hours correcting my English, pointing out flaws in my logic, and supporting me, as she has done for the past 67 years.

My interest in Artificial Intelligence (AI) goes back a long way, but my research career didn't start with AI. After graduating in Electrical Engineering in 1948 at Imperial College, I joined John Lamb's research group studying fast chemical equilibria in liquids using ultrasonic waves, and, with a PhD, continued the research in ICI's Akers Research Laboratories, Old Welwyn, North of London. My publications from that era have been cited more than my AI research, but ICI suddenly closed the laboratories in 1961. Eric Ash, a fellow student from undergraduate days, found a job for me with Standard Telecommunication Laboratories in Harlow, Essex. From a list of topics offered by Len Lewin, I was unable to resist the appeal of *The Electronic Simulation of Cerebral Functions*.

My first learning machine was published in 1963 and called STeLLA after the name of the laboratories. Peter Joyce built STeLLA, using post office relays because transistors weren't yet generally available and vacuum tubes were too large. STeLLA moved around the laboratory floor, but the relays were too unreliable and so we had to resort to computer simulation. Brian Gaines deepened my understanding of control systems theory and stochastic computing.

With Molly's family being in Dunedin and my family having moved from India to Matamata, it was an easy decision to move to the University of Canterbury in Christchurch, New Zealand, when the opportunity arose in 1966.

Work on STeLLA continued with an attempt by Peter Cashin to make it powerful enough to handle language, but it was the mathematical ideas of John Cleary that inspired my first version of PURR-PUSS, now called PP, and my first book. Igor Aleksander enabled publication of my second book. Bruce MacDonald, Shaun Ryan, and Kon Kuiper contributed to the development of PP as recorded here and in my second book.

I am grateful to Natalie Farra of Elsevier for guiding this book through the stages of publication at a difficult time.

Many thanks to everyone who helped me with this book and in my research.

Note. My daughter Gillian Andreae's robot cat illustrations first appeared, with her permission, in my "Man-Machine Studies" reports, 1972-1991, ISSN 0110 1188, then in my two books "Thinking with the Teachable Machine" and "Associative Learning for a Robot Intelligence", and more recently on her commercial items.

JohnHAndreae@gmail.com

Chapter 1

Brain, Body, and World

Introduction

The robots are ready. All they need now is a brain.

New Scientist (Cover, 2019).

This book is about a brain for a robot which has a body like ours. The aim is for the brain to make the robot behave like us. Neuroscience has discovered a lot about our brains but not yet enough to tell us how to design a working brain for a robot.

There is a common belief among Artificial Intelligence (AI) researchers that the human mind is a computer program in the brain. For example, Eric Baum wrote:

I believe evolution, using an amazing amount of computational power, produced an amazingly compact program capable of exploiting the structure of the world. ... The mind is a computer program.[1]

This can't be right. Evolution develops living things, animals, and plants, by experimenting with huge populations of individuals over long periods of time. Each individual tests the mutations it has been given. Successful mutations give individuals a greater chance of passing on their genes to their progeny for further testing. The human body hasn't changed much over hundreds of thousands of years, so we can expect evolution, over that time, to have gradually developed the human brain to make the best use of that constant human body. If the human body had been changing, evolution would have had little time to develop the best way to use the most recent version of the body.

The world that humans live in, unlike the human body, changes from generation to generation.[2] Humans lived through climate changes that altered the environment in major ways. They spread across the world into different geographical regions with mountains, lakes, rivers, forests, savannahs, and deserts. More recently, humans themselves have changed

An AGI Brain for a Robot. https://doi.org/10.1016/B978-0-323-85254-8.00001-6
© 2021 Elsevier Inc. All rights reserved.

the world by agriculture, industry, and science. During a human's lifetime, new experiences are encountered continually. The brain of a human baby meets a new world with a body similar to those of its ancestors.

It is reasonable to expect evolution to have equipped the human brain with programs to make the best use of the sensory-motor equipment provided by the unchanging human body.[3] Programs convert the output of the cells in the eyes into high level information which the central brain needs in order to recognize objects and faces and to follow movement and changing shape. Other programs convert the output of the auditory nerves into meaningful sounds, rhythms, and the elements of words. Programs convert high level commands from the central brain into the details of muscular movements. Following Fodor in his book *The Modularity of Mind*, I will call these programs **modules**.[4]

It is not reasonable to believe that evolution has provided the human brain with a program that tells it how and what to do in an ever-changing world. In fact, we know that evolution discovered a way to enable humans to learn from the world they were born in. Infants, children, and adults can be seen to learn. Humans learn from other humans by reinforcement learning (reward and punishment)[5] and by imitation; they also learn by exploration and, after infancy, by being taught. The program that enables a human to learn must be like the operating system of a computer, or like a word processor. The operating system enables a programmer to write any program without telling the programmer what to do. The word processor enables the writer to write any text without telling the writer what to do.[6] The program that enables the central brain to learn must organize the learning without telling the brain what to do.

In this book, I describe a brain for a robot to enable the robot to learn like a human. The robot is called **PP**,[7] its brain is called the **PP brain** and the program that organizes the learning in the PP brain is called the **PP program.**[8]

The PP brain learns like an infant, step by step, and a little at a time. I show how the PP brain, in a robot body, can interact with the world, can begin to learn language, use Working Memory, create its own goals, have free will, plan, and, perhaps, even become conscious.

The Brain is like a River

The PP brain is like a river.[9] No designer tells the river where to go; no programmer tells the PP brain what to do. The river is driven by gravity; the PP brain is driven by expectation. The river's course is determined by the terrain it flows over; PP's behaviour is determined by its experience. The river flow is affected by solid rock and soft soil, by waterproof clay and permeable gravel, and by cracks and crevices; the PP brain's decisions are affected by what has happened to it in the past, by what it has done before, by what it has seen others do, and by novelty. The river can flow through man-made channels and will flow around or over dams that block its way; the PP brain seeks approval and avoids disapproval. The river consists of water, which is added by streams and rainfall; the PP brain consists of associations, which are added by experience.

The PP brain avoids the limitations of programs established by Gödel and Turing, by not being a formal program.[10] You may have difficulty in persuading or teaching PP to do some task that you have in your mind, but that difficulty will not be due to errors that someone has programmed into the PP brain.

The PP brain needs a body through which it can interact with the real world. *Sometimes people think of their bodies as cages within which they live, but bodies are quite the opposite of cages. My body connects me to the world around me in such an intimate and interactive way that I become part of that world. I can participate in the dynamic society and culture of humanity. My cage is wide open.*[11]

The associations in the PP brain, which take the place of a top-level program controlling the robot body, are acquired by the interaction of the robot with the real world.

Not having a real robot body for the PP brain, we have to be content in this book with small simulated worlds and a real teacher (me). In chapter 3, the PP brain will become the brain of one robot in a simple simulated world, while I, as teacher, am the brain of a second robot in the same world. Both robots can move about, talk, and point with their hands. Hopefully, someone will put the PP brain into a real robot body some day soon.

Simplifying the Human Brain

Many researchers agree that the way most people imagine AI – a machine that thinks just like a human – is a remote prospect, unlikely to be fulfilled without a better understanding of how our own minds work. There is a broad consensus that such artificial general intelligence (AGI) is achievable this century, but few believe it will result from just carrying on as we have so far.

New Scientist Instant Expert (Heaven, 2017)

... human-level AGI isn't within sight. (Boden, 2016)

The moral component, the cultural component, the element of free will---all make the task of creating an AGI fundamentally different from any other programming task. It's much more akin to raising a child. (Deutsch, 2019)

Despite the remarkable advances in computing, the hype about AGI---a general intelligence machine that will think like a human and possibly develop consciousness---smacks of science fiction to me, partly because we don't understand the brain at that level of detail. (Ramakrishnan, 2019)

The structure and operation of the human brain is so complex that even the specialists don't fully understand it,[12] so how can we talk sensibly about its intelligence, free will, and consciousness? The PP brain is a 'bare bones' working brain that is stripped of all the extra complexity, so we can see what is really needed to have those characteristics. PP has free will and can control its body, learn, plan, and use language, but PP has not been given emotions. The richness of vision has also been deferred until the brain can be put into a real robot so that the world doesn't have to be simulated. Inevitably, much of the magic of the human brain is missing until the robot has good vision, but in this book, the focus is on language.

I start by showing how modules feed information from the sensors of the robot body to the PP brain and how more modules accept commands from the PP brain to control the effectors (muscles) of the robot body. Then I explain how the PP brain takes into account the effects of the context of its actions and predictions by means of Groups of associations. Goal-seeking with LeakBack gives the PP brain motivation. Novelty goals give the PP brain free will. Planning sees the PP brain wandering through its memory of past experience and discovering new possibilities. Beginning to learn language using Working Memory is demonstrated by an interaction between two robots, one controlled by the PP brain and the other by a teacher (me). The last chapter points the way to successfully test the PP brain for consciousness in a real robot.

As mentioned above, PP uses **modules** to transform information from the senses (vision, sound, smell, taste, touch, proprioception, etc.) into high level information (stimulus events) for the PP brain. (Proprioception is sensory information about what the robot's effectors or muscles are doing.) By 'high level' I mean more structured than low level. Instead of the individual tones in a speech wave, which are low level, the PP brain is given Words that are high level. Instead of pixels from the retinas of the eyes, the PP brain is given information about the objects seen.

Each module outputs stimulus **events** of a different **event-type** (type in the sense of category). The PP brain is given one event of the event-type of each module in each step. There are also modules that are 'motor programs' which take high level commands (action events) from the PP brain and convert them to muscle movements of the body. Each motor program module handles one action event-type and takes one action of that event-type from the PP brain on each step (see Figure 1).

The high level stimulus events come via modules from a variety of sensors: eyes, ears, nose, tongue, touch sensors, movement sensors, pain receptors, and various sensors internal to the body. There will be one or more modules for each sensory mode and one event-type from each module.[13] Thus, when PP can be provided with good vision, we can expect event-types for colour, lines, shapes, textures, faces, etc. according to what is then known about the visual information fed into our brains. There will be many stimulus events over time for each event-type. For example, the colour event-type will have stimulus events of red, blue, green, yellow, mauve, etc. An event-type for words would have different words as events. When someone speaks we hear words, not sound waves. It is rather unlikely that there would be an event-type for words in the human brain, but in the simple experiments I describe later, I do assume an event-type for Words just to sidestep the segmentation of sound streams into words.

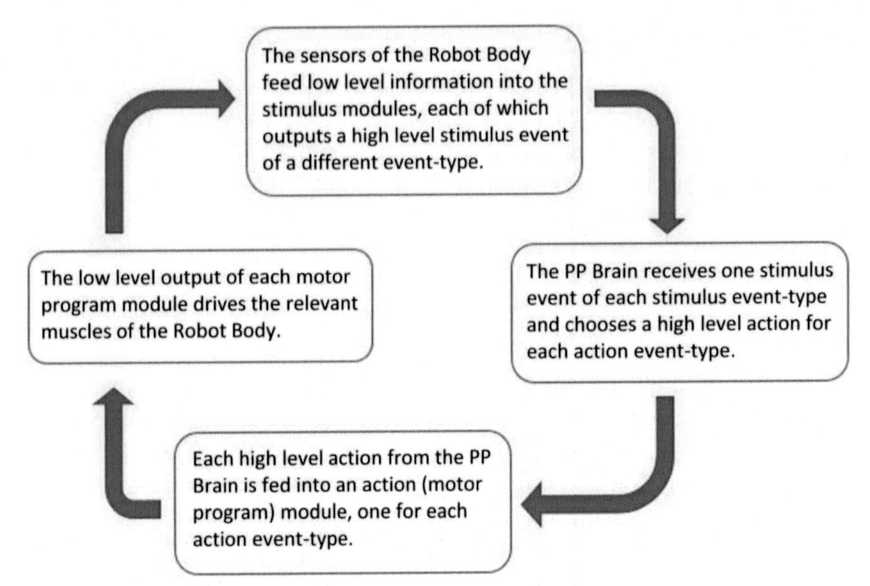

Figure 1 One Step: Events, Event-types, and Modules.

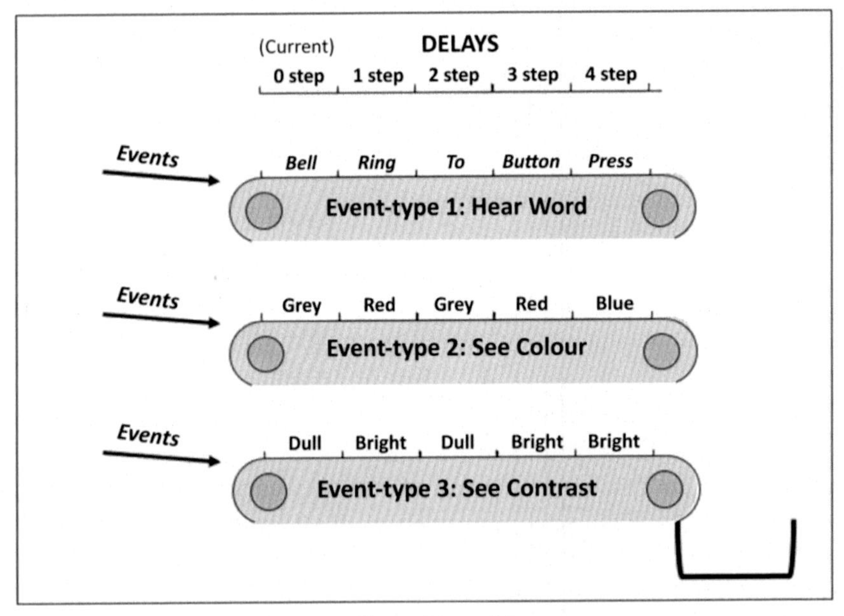

Figure 2 Short Term Memory.

The recent successes of AI with Deep Learning[14] can be seen as similar to the evolutionary process by using huge streams of data fed into a multi-level neural network. Thus, a module for the segmentation of speech into words could probably be achieved by Deep Learning.[15]

Here is an example of an event-type and an event from chapter 3: BodyMove is an action event-type, FORWARD is an event of that event-type, and the effect on the robot body is to move it forward.

The brains of mammals have evolved through the millennia. From the earliest times, axons of neurons carrying stimulus events from the outputs of different modules will have travelled into the area of the cerebral cortex where associations are formed. As the brains evolved, modules will have multiplied into more modules for more event-types. An important development would have been the production of **delayed event-types**, needed for associations that could support sequences of movements and sounds, the harbingers of language, both sign and vocal. Delayed event-types make an important contribution to PP's ability to learn language by holding the last few Words that it says or hears.[16]

Short Term Memory

Short Term Memory is where all the events from the modules are collected at each step of the operation of the PP brain. It can be thought of as a number of conveyor belts (first in, first out buffers), one for each event-type with its delayed event-types (see Figure 2).

Each conveyor belt is divided into slots and the belt moves forward to the right by one slot at each step of the operation of the PP program (for that event-type). Events are taken onto the conveyor belt for their event-type and, as each event moves onto the front slot (current: zero delay), the previously current event moves forward a slot (1 step delay), pushing the earlier events in the more delayed slots even further to the right. The events in the oldest slots on the belts drop off the end at the right and are lost. The conveyor belts are just long enough to hold events with the longest delay required. An experiment[17] showed that Miller's famous magic number 7 ± 2 is about right for the maximum number of delayed steps. All the conveyor belts used in the PP for chapters 3–5 are of length 8.

Many people see the human brain as having a Global Workspace for the centre of consciousness.[18] PP's Short Term Memory is like a Global Workspace but when we come to planning later in the next chapter, there will be a need for a second Short Term Memory, STM-plan. That will pose a problem for understanding Short Term Memory as a centre of consciousness.

Up to now,[19] PP has had very simple robot bodies and worlds simulated on a computer. These simulations have checked the validity of PP from several points of view. In chapters 4 and 5, PP will be moving about, talking, and Pointing with its Hand in a small simulated world. All of PP's actions will be chosen by the associations it learns. When someone with the resources can put the PP brain into the head of a real robot, it will become possible to have it interact properly with the real world.

Multiple Context Associative Learning

Much of human learning is based on observation and mimicry, and we need far fewer examples than deep learning to learn to recognize a new object. Unlabelled sensory data are abundant, and powerful unsupervised learning algorithms might use these data to advantage before any supervision takes place. ... Unsupervised learning is the next frontier in machine learning. We are just starting to understand brain-style computing. (Sejnowski, 2018)

Until we solve the basic paradox of learning, the best artificial intelligences will be unable to compete with the average human four-year-old. (Gopnik, 2019)

The classical example of associative learning was Pavlov's dog learning to associate the expectation of food with a single stimulus, the sound of a bell. The PP brain learns by associating actions and stimuli with multiple stimuli, called contexts. This is Multiple Context Associative Learning.[20]

The PP brain learns incrementally, bit by bit, by exploration (seeking novelty), imitation, and approval-seeking. The PP brain can learn from single exposures to actions and stimuli, which is difficult for statistical approaches like Deep Learning. Learning the sounds of one's language, learning to segment speech, and learning to recognize patterns, such as faces, require statistical learning, like Deep Learning, so they have been side-stepped in this book.

The PP program doesn't tell PP's robot body what to do, but it does organize how the PP brain learns from its experience with the real world. The PP robot is controlled by the PP brain's networks of **associations** acquired by interacting with the real world. The associations are in control.

The PP program is like the operating system of a computer, but, instead of allowing programmers to put programs into PP's memory, it allows the real world to put experience into PP's memory.

In chapters 4 and 5, all the actions chosen by the PP brain for its robot body will be chosen by the associations that the PP brain learns while interacting with the second robot, controlled by me, in the simulated world.

We act when the **context** is right for the action.[21] To pick up a cup, we usually see first that the cup is within the reach of our hand. The context is right for picking up the cup. To put the cup on a shelf, we check that it is the right shelf, that there is space on the shelf, and that the cup feels secure in our hand. In the same way, we predict what is going to happen from the current context. If there are dark grey clouds in the sky, we expect rain. If there is blue sky and we have just heard a favourable weather forecast, we expect the rest of the day to be fine.

The associations learned by the PP brain associate context with an action event, or context with a stimulus event. Associations of context with action events tell the PP brain what to do in a particular context. Associations of context with stimulus events tell the PP brain what the body and world might do next in the current context. The action event or stimulus event linked to the context is called the **associated event** or the **predicted event**.

At any moment in time there is an enormous amount of context which could be considered and a lot of it will be irrelevant for the actions or predictions we need to make. Just think of everything that you can see, hear, touch, taste, and smell, not to mention your internal feelings. The PP program uses Groups to allow many combinations of <u>event-types</u> as contexts, without restricting the <u>events</u> that can belong to those event-types. The origin and operation of the Groups are explained in the next chapter.

PP has its Own Goals

At birth, the human brain is set up to be attracted to novelty. (Berk, 2013)

It is no mere coincidence that the philosophical problems of consciousness and free will are, together, the most intensely debated and (to some thinkers) ineluctably mysterious phenomena of all. (Dennett, 2018)

I can do what I want. But I can't want what I want, so there are deep inbuilt constraints. (Velmans, 2006)

The associations that the PP brain learns are stored in its Long Term Memory. Some of the associations have their associated events marked as goals because the decisions they represent led to automatic approval or to approval given by teachers. If the PP brain had all the associations it had learned from its experience with the real world in its Long Term Memory, and it used them only for reaching goals which were given to it automatically or by a teacher, then PP would lack the initiative and creativity characteristic of a human. It would be just an obedient robot. It is essential that PP makes its own goals and seeks them.

Novelty goals are the PP brain's own goals. Every new association stored in the PP brain's Long Term Memory has its associated event marked as a goal. When that association occurs again, the goal mark is removed. PP tries to reach its marked novelty goals. These goals are part of the basic associative process of the PP brain and are not goals that PP is aware of, because they are not part of the PP brain's sensory information in Short Term Memory.

An association is new because its particular combination of context events and associated event is occurring for the first time. PP might be seeing an object for the first time or it might be just a difference in colour or scent. The events are determined by the robot's body and the real world it is in. The association is new because it is not already stored in the PP brain's Long Term Memory. Therefore a novelty goal is determined by what is in the PP brain's Long Term Memory and by the real world acting through the robot's body. In a real sense, PP, as an individual agent, is the content of its Long Term Memory. PP acts, thinks, and plans by means of the associations in its Long Term Memory. PP's novelty goals are novel to it alone.

PP can also be given goals by approval (like a reward) from a teacher and by automatic approval, but these are not its own goals. Automatic disapproval (like pain) from the robot body and disapproval from a teacher set up negative goals to be avoided. The approval goals normally take preference over PP's novelty goals, and when PP is seeking approval goals its behaviour is being influenced by the teacher.

In chapter 4, the PP brain in its simulated robot is mostly seeking the approval goals given by me through the second robot, because I am trying to teach PP quickly. In chapter 5, I give no approval, so PP is seeking its own goals and a few goals from the automatic approval it gets from 'Squashing the Cake' in the simulated world.

Novelty goals make exploration and play natural behaviours of PP.

When PP is seeking its own novelty goals, it is "doing its own thing" and can be said to have a basic level of "free will", akin to that of an animal like a cat. To achieve human-level free will, PP needs to learn language and be able to think about, and talk about, what it wants to do. It is not difficult to imagine how PP, when it has learned to use language and to reason, will be able to **generate its own novelty goals** by thinking of new ideas representative of its desires, such as watching humans play football or going for a sail in a yacht. In this way, PP could create new associations with novelty goals and become aware of choosing actions because of its desires.

Douglas Hofstadter,[22] in his book *I Am A Strange Loop,* has difficulty with the idea of free will:

What, then, is all the fuss about "free will" about? Why do so many people insist on the grandiose adjective, often even finding in it humanity's crowning glory? What does it gain us, or rather, what would it gain us if the word "free" were accurate? I honestly do not know. I don't see any room in this complex world for my will to be 'free'.

He seems to see it as an ability not to do what one wants to do: a freedom from one's desires. In PP, free will is a freedom from built-in desires and a freedom from wanting rewards given by other agents. This is a freedom that comes from having its own novelty goals. I am happy with that.

Whether, or not, the world is deterministic makes no difference to free will: our need to make the best choices to achieve the best outcomes remains the same. Daniel Dennett[23] in his classic book on free will, *Elbow Room,* seems to me to imagine PP:

Yes, if we try hard, we can imagine a being that listens to the voice of reason and yet is not exempted from the causal milieu. Yes, we can imagine a being whose every decision is caused by the interaction of features of its current state and features of its environment over which it has no control --- and yet which is in itself in control, *and not being controlled by that omnipresent and omnicausal environment. Yes, we can imagine a process of self-creation that starts with a non-responsible agent and builds, gradually, to an agent responsible for its own character. Yes, we can imagine a rational* and deterministic *being who is not deluded when it views its future as open and "up to" it. Yes, we can imagine a responsible, free agent of whom it is true that whenever it has acted in the past, it could not have acted otherwise.*

There is another important feature of finding and removing novelty goals, which I mentioned in my 1977 book, *Thinking with the Teachable Machine*. A new association hasn't any forward network connections in Long Term Memory because it hasn't "been there" before. When the PP brain reaches the same novelty goal again and removes the novelty goal mark, the PP brain has made a path from the new association through Long Term Memory and back to the new association. The new association has been integrated into the PP brain's Long Term Memory.

When the PP brain is making decisions on the basis of the experience it has gained in the real world and its goals, it is easy to see that the PP program is just a facilitator and that the PP brain has become its own store of experience with goals. The PP brain is in control.

We start the PP brain with no associations in its Long Term Memory.

This is called a ***blank slate*** or ***tabula rasa*** start and it has the advantage of making clear what has been learned: everything in Long Term Memory has been learned. Of course, we start the PP robot body with modules for processing sensory data into high level stimulus information for the PP brain and with motor program modules to accept high level commands (actions) from the PP brain to operate the robot's 'muscles'. These are the modules mentioned earlier. Also, the robot body is given reflex movements which move it around the world until the PP brain has learned enough to take over. PP has plenty of innate or built-in processes.

Aside-1. John Locke assumed a blank slate in 1690 more than 3 centuries ago: *Let us then suppose the mind to be, as we say, white paper void of all characters, without any ideas. How comes it to be furnished? Whence comes it by that vast store which the busy and boundless fancy of man has painted on it with an almost endless variety? Whence has it all the materials of reason and knowledge? To this I answer, in one word, from EXPERIENCE. In that all our knowledge is founded; and from that it ultimately derives itself. Our observation, employed either about external sensible objects or about the internal operations of our minds perceived and reflected on by ourselves, is that which supplies our understandings with all the materials of thinking.*[24]

Being controlled by its own associations and seeking its own novelty goals are the two features which characterize PP and make it like us. Much research lies ahead to give this combination 'intelligence'. In this little book, I try to show you what has been achieved and how much more remains to be done. In chapter 4, I teach PP to use Working Memory, which is essential for language and most kinds of thinking. Working Memory is something that Steven Pinker said was impossible for associative learning to implement.[25]

Aside-2. *Yet I assert that the only general principles, which associate ideas, are resemblance, contiguity, and causation.* So said David Hume in 1740. It is not difficult to relate these three principles to the associations that I have defined 280 years later!

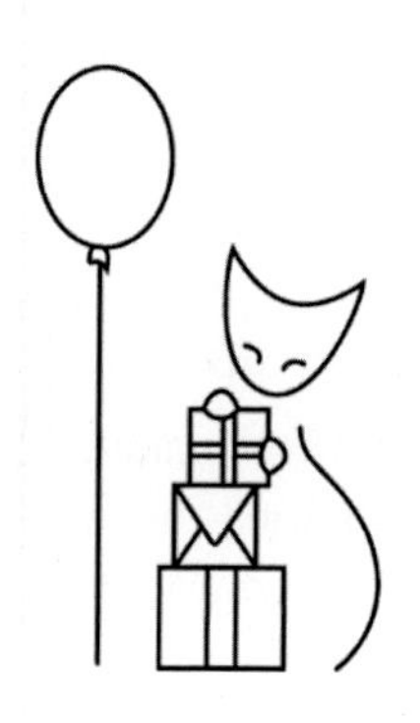

Chapter 2

Groups

I feel that we're just really exchanging anecdotes with each other; no one has done what they've done for physical sciences: come up with a proper framework that everyone buys into, with laws and rules and principles and so on, that successfully brings together the different levels of working in the brain.

(Greenfield, 2006)

Different areas of the human cerebral cortex are involved in different cognitive and behavioural functions. For example, there are areas for spatial awareness, word formation, language comprehension, and so on. This suggests that a particular area of the cortex brings together events of event-types relevant to the function of that area.[26] A baby is born with these areas, which have evolved over thousands of years to match the brain to the facilities of the human body. I am not saying that every baby is born with the same connections, but we can expect their brains to start with similar connections. A **Group** is a simple equivalent to one of these areas, describing what event-types (for sight, sound, touch, taste, smell, and movements of head, arms, hands, legs, etc.) are input to the area to form contexts, and what associated event-type (e.g. BodyMove action, visual stimulus) is output from the area.

Any innately specified collection of different event-types for a context and associated event-type in PP is called a Group. These Groups are used to form associations that enable the PP brain to predict actions for its own decisions and to predict stimuli for making its plans.[27] A Group specifies that the contexts of its associations include events from particular slots on the conveyor belts of Short Term Memory.[28]

We know very little about the information fed into the human cerebral cortex, but we do know that the raw information from sensors like the eyes and ears is processed into higher level information before being used in the cortex. Similarly, in the case of the PP robot, we assume that the event-types assigned to Groups are not raw sensory and motor information but information processed from the raw data to be as useful

An AGI Brain for a Robot. https://doi.org/10.1016/B978-0-323-85254-8.00002-8
© 2021 Elsevier Inc. All rights reserved.

as possible. This is done by the built-in (innate) modules. (The 'blank slate' start is only for Long Term Memory and its associations.)

The range of Groups and modules used by the PP program should be seen as designed to make the best use of the sensors and effectors of the robot body. As in the case of the human cerebral cortex, it is not possible to provide all of the near-infinite contexts of events available from sensors and effectors. A Group describes what event-types from the modules can be brought together to make an association. The robot body and the real world determine what actual events form associations under a particular Group. The Groups give PP its computational strength.

The Groups may be likened to a carpenter's bag of tools, the use of those tools being the associations made possible by the Groups. As long as carpenters have tools that cover a wide range of jobs, they are well-equipped. Carpenters cannot keep tools in their bags for every specialized task. A brain cannot be given all possible Groups. In my early work on computers with less than a megabyte of memory, PP had to be tested with just a few Groups and so my interest was in the choice of good Groups. Now that the amount of computer memory is not a serious problem, I can be generous with Groups. They need to cover the context space for a wide range of tasks.

For many tasks, like speaking, manipulating the hands, and visual scanning, sequential information is important. Groups allow for this by having delayed event-types. A Group might have the last three Speech events as part of its context description. Details are avoided as much as possible in this book so that it is accessible to the general reader, but it is difficult to understand Groups without an example. Here is how a simple Group GA could be described:

[GA: HearWord(-3), HearWord (-2), HearWord (-1), Eye(-1), Touch(-1) ► Move-Hand(0)]

And here are two associations, 1GA and 2GA, that could be formed and stored in the PP brain's Long Term Memory using Group GA:

[1GA: *Press, The, Button,* See-Button, Not-Touching-Button ► Move-Hand-Down]
[2GA: *Release, The, Button,* See-Button, Touching-Button ► Move-Hand-Up]

The arrow symbol ► can be read as "predicts" or "is associated with". There can be one or more event-types, or events, in the context to the left of the ► sign and they can be listed in any order.

The context of a Group is a collection of event-types; the context of an association is a collection of events.

In the Group GA, the HearWord event-types have delays of 3, 2, and 1 steps, indicated by (-3), (-2), and (-1). Both the event-type from the Eye and that from Touch have one step delay (-1). The event ***Press*** in association 1GA is of event-type HearWord(-3) in Group GA. ***Press*** has a delay of 3 steps because it came before ***The*** which had a delay of 2 steps, which came before ***Button*** which had a delay of 1 step, which came before Move-Hand-Down that occurred on the step that the association is stored (no delay). The event ***The*** in both associations is of event-type HearWord(-2) in Group GA. The Move-Hand event-type is the **associated event-type** (also called the **predicted event-type** because it is used for prediction) with zero delay (i.e. now) and it follows the **context event-types of the Group**, which is the collection of event-types to the left of the ▶ symbol. Similarly, the associated event (also called the predicted event) Move-Hand-Down in association 1GA is of event-type Move-Hand(0) in Group GA.

The association 1GA is like a rule saying "If you hear ***Press The Button,*** are **seeing the button,** and are **not touching the button** then **move your hand down** to press the button". However, it is best <u>not</u> seen as a rule that <u>must</u> be obeyed, because there will be other associations competing with it.

The Groups are given priorities. If there is more than one Group with the same associated event-type, associations of a higher priority Group are given preference over those of a lower priority Group. A Group with more event-types in its context usually has a higher priority than a Group with fewer event-types because the larger context is more **specific**: more events in the Group's associations have to match the arriving events. Small Groups allow the PP brain to **generalize** because fewer events are specified by their associations. If the PP brain had another Group GB:

[GB: HearWord (-1), Touch(-1) ▶ Move-Hand(0)]

with associations

[1GB: *Press,* Not-Touching-anything ▶ Move-Hand-Down]

[2GB: *Release,* Touching-something ▶ Move-Hand-Up]

they would allow a generalization of the pressing and touching to any object, whether it was a button or not. If it were a button, we would want the associations of GA to be used because GA is more specific than GB. The more specific GA would be given a higher priority than the more general GB. Some Groups with quite different event-types in their contexts may not be obviously more specific than others and would be given the same priority. Usually, if one Group has more event-types than another, it is given a higher priority because it is more specific.

If Groups and their priorities are used in the human brain, they will have been selected by evolution. In the case of PP, they have to be chosen by the designer of the PP program. The PP program is simple.

The choice of Groups is not difficult. The difficult part of putting the PP brain into a real robot will be the designing of modules that make the best use of the robot's sensors and effectors.

It is easy to imagine associations being biologically implemented as neurons, with dendrites as inputs of the context and axons as outputs of the associated events. Figure 3 shows what association 1GA would look like.

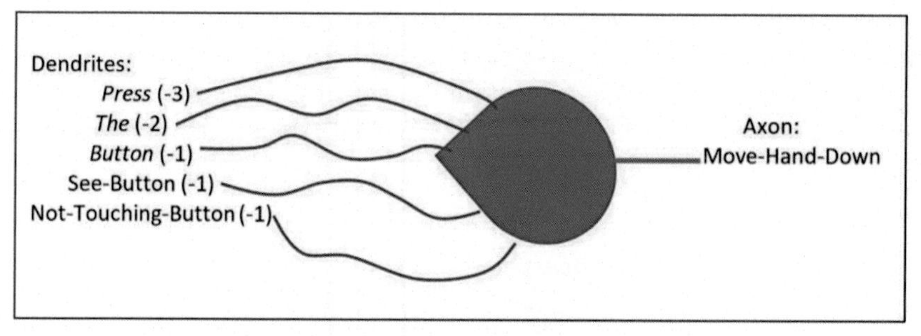

Figure 3 Neuron for Association 1GA.

The Evolution of Groups

Darwin always stressed the continuity of life, how all living things are descended from a common ancestor, and how we are in this sense all related to each other. So humans are related not only to apes and other animals but to plants too. The River of Consciousness. **(Sacks, 2017)**

Groups should be seen as enabling rather than limiting, as selecting rather than preventing. The human brain may have a way of organizing its associations that is different from using Groups of event-types, but it still has to select combinations of events for the associations it stores in Long Term Memory.[29] Grouping together associations that have the same combination of event-types into a single Group allows us to assign different priorities to Groups and, for Groups with associated event-types that are actions, we can construct networks that link the associations together.

As animal bodies evolved, new event-types will have become available and Groups will have evolved to make the best use of those event-types. For instance, at one extreme, the jelly-fish has a simple nervous system and just reacts to the world around it. At the other extreme, humans evolved with a vocal apparatus that gave them the wider variety of sounds needed for language. We can guess that Groups with more delayed event-types will have evolved at the same time to support the sequential nature of language.

Walking is a repeated, sequential behaviour, but to learn to walk PP wouldn't need delayed event-types. Each movement changes the position of the robot's limbs, and the new position, together with the rest of the context, tells the robot what movement to do next. Language, both spoken and signed, is also sequential, but each word or sign doesn't say what the next word will be, so it is helpful to have delayed event-types which can hold the past few words or signs.

The number of Groups in the human brain can only be a tiny fraction of the number of possible Groups. This is because, even with conservative assumptions about the numbers of event-types and how many can be combined into a Group, the number of possible Groups exceeds the number of neurons in the cerebral cortex (nearly 100 billion).[30] Then, if every neuron in the cerebral cortex implemented an association, there still wouldn't be enough associations to populate the possible Groups with one association each.

Another way of seeing this, is to consider all the possible Groups that could be defined using the conveyor belt slots of Short Term Memory. A Group is a selection of event-types (positions of slots on conveyor belts). In the human brain this choice would be made by evolution, if this is how the brain works. For a robot, the designer of the PP program must make the choice.

The brain is plastic in many ways,[31] so there may be a need for the addition of new Groups after long periods of interaction or for repair to Groups because of damage.[32] Experiments on how to do this addition or repair will have to wait until very long interactions become possible with the PP brain in a real robot.

A good reason for expecting the Groups provided at birth to remain the same throughout life (unless damage requires repairs, using stem cells, perhaps) is that the associations stored within a Group are connected to each other by the networks to be described shortly. Also, if there is learning at two levels that depend on each other, the learning at one level must take place on a different time scale from the learning at the other level to avoid confusion.[33] This suggests that the selection of Groups is the job of evolution or robot design while the learning of associations should be lifetime learning. Also, a brain can't afford in its short lifetime the trial and error needed for finding new Groups, which evolution, by mutation and natural selection over large populations, can indulge in.

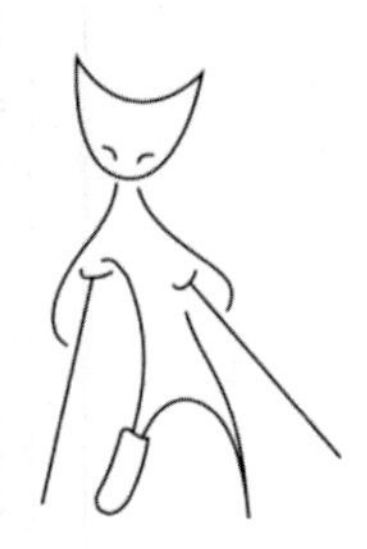

A Fairy Tale about Groups

Let's imagine how a new human brain might grow. We don't know how the brain works and I don't have neurological explanations for how any of the following steps could be achieved, but I am thinking of the kind of biological model that would be equivalent to PP. Readers may find it helpful to have this picture of neurological Groups in the back of their minds.

1. Not more than 100 billion neurons are distributed throughout the main part of the cerebral cortex where associations will be formed. We assume that each neuron, in what is here called the **association cortex**, is capable of forming an association.

2. Neural circuits grow outside the association cortex to form the modules which construct event-types from raw sensory information. Each module has a bunch of output neurons which, by their activity, represent events of its event-type.

3. Neural circuits grow in the motor cortex to form motor programs that will drive the muscles of the body. The motor programs will be driven by axons from association neurons or, if none are active, by axons from neurons in reflex circuits.

4. Axons from neurons in the association cortex grow into the motor cortex where they "await instructions" for connecting to appropriate motor program neurons when the latter are activated by other association neurons or reflexes.

5. **Bunches** of axons from output neurons of each of the stimulus event-type modules grow into the association cortex. Their paths through the association cortex are roughly specified innately so that bunches of axons from different modules come close to each other as often as possible, and so that areas of association cortex correspond roughly to major functions of the brain such as language, manipulation, and problem solving. A Group is formed wherever a number of bunches of axons are close enough to form a context of events for neighbouring association neurons within dendritic range.

6. An association neuron that is within dendritic range of a Group, when axons of that Group are active, competes with neighbouring association neurons to form an association and have its axon activated in the motor cortex.

The main idea of this picture is that Groups are formed where bunches of neurons representing different event-types come close together. As evolution provided animals with more and better sensors, more and better context information became available. If the PP design is relevant, evolution then provided the animal brains with more and better Groups. Eventually, humans arrived and were given large Groups with enough delayed event-types to support human language and advanced thought.[34]

A Warning!

The event-types and events in Groups are given helpful names for teacher's, that is my, benefit. They mean nothing to PP. In the simple World described in chapter 3, the objects are called Cake, Stump, and Wall. The actions of a robot are called FORWARD, RIGHT, LEFT, and WAIT. These names help me, the teacher, but mean nothing to PP when it starts to learn. When we interpret PP's responses to teaching, we will treat these names as nonsense Words to avoid ascribing more intelligence to PP than it deserves. Failing to treat these names as nonsense words is called **encodingism.**[35]

Avoiding encodingism, by using nonsense words, is one of two methods I use to separate what PP has really learned from what it appears to have learned. The other method is to look at the associations in PP's Long Term Memory.

Networks in Long Term Memory

Each Group with an associated event-type that is an action has a network showing which associations of the Group follow which.[36] The network helps the PP brain to choose actions which lead it to its goals. Groups with associated event-types that are stimuli are not given networks. Stimuli come from the robot's body and the real world so they can't be chosen by the PP brain, and a network wouldn't help it.

A network is defined as a collection of nodes joined together by transitions.

The same context of events prescribed by a Group can be in associations with different associated events, so it is convenient to make each different context of events a **node** of the network of a Group. Even though stimulus-predicting associations don't have networks, it is convenient to call each different context a node in their case too. Here is an example of two associations with the same context (node) but different associated events:

At the beginning of this chapter, we had an association of Group GA:

[1GA: *Press, The, Button,* See-Button, Not-Touching-Button ▶ Move-Hand-Down]

Now suppose that another event of event-type Move-Hand is Move-Finger-Down, and that we have an association using it:

[3GA: *Press, The, Button,* See-Button, Not-Touching-Button ▶ Move-Finger-Down]

In a network of Group GA, associations 1GA and 3GA are represented by <u>one</u> node for the common context (*Press, The, Button,* See-Button, Not-Touching-Button).

The **transitions** of a node have two jobs. First, they have to say which associated event of the associations having their context in a node is in the transition, and second, they have to say which nodes might follow that associated event in PP's step-by-step behaviour.

Figure 4 shows an example of a small part of a network. The broken arrows suggest transitions to and from the rest of the network for this Group.

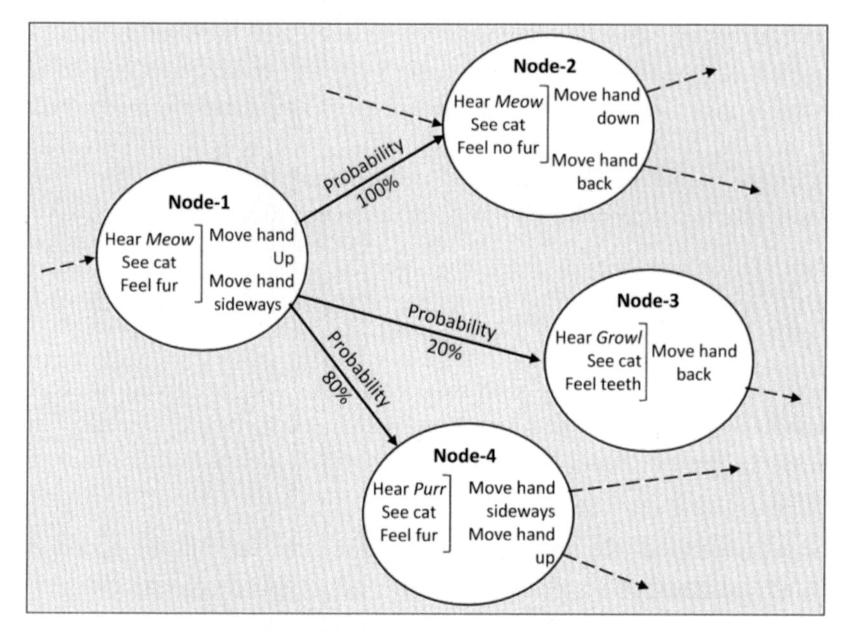

Figure 4 Part of a Network.

Node 1 has a context with two different associated action events:

[Node-1: Hear *Meow,* See cat, Feel fur ▶ Move hand up, Move hand sideways]

Node-1 is equivalent to two associations:

[Association-1: Hear *Meow,* See cat, Feel fur ▶ Move hand up]

[Association-2: Hear *Meow,* See cat, Feel fur ▶ Move hand sideways]

The context, **Hear *Meow,* See cat, Feel fur,** points directly to where the node is stored in Long Term Memory.

The associated events are in a list attached to the node, the most recent associated event, **Move hand up,** being at the head of the list. Every time in the past that **Move hand up** has followed the context, the next node has been Node-2, so the transition probability is 100%. When the other associated event on the list, **Move hand sideways,** has followed the context, Node-4 has followed 4 times as often as Node-3 and so the transition probabilities are 80% and 20%.

The importance of having the most recent associated event at the head of the list[37] can be illustrated by supposing that the context referred to my driving our car and looking at the petrol gauge, with the associated event saying how much petrol was in the tank each time I looked. Most significant would be the associated event at the head of the list, which would tell me how much petrol was there the last time I looked.

Associated events can be marked as novelty goals or approval goals or with disapproval. Associated events and nodes are given expectations by LeakBack calculations that we come to shortly.

Long Term Memory is the sum total of all the associations in networks with their nodes, associated action events, and transitions, together with the associations in nodes of Groups that have associated events which are stimuli.

To summarize the relationships of association, associated event, and node:

- an association is a context of events plus an associated event;

- a node is all those associations of a Group that have the same context.

It is not quite correct to say that associations are stored in the PP brain's Long Term Memory. In actual fact only the associated events and transitions are stored. The context of an association is an address which points to where in Long Term Memory the node with the associated event is stored. Most of the time, however, it is easier to think of the whole of the association as being stored and I will continue to say that associations are stored in Long Term Memory.

Long Term Memory holds PP's Experience

Long Term Memory is the sum total of PP's experience. Think of it as layers of networks, one network for each Group and each layer having Groups of the same priority. The layers of higher priority Groups are above layers of lower priority ones. Higher priority Groups usually have more event-types than lower priority Groups, but there will be exceptions where some event-types are much more significant than others. In human brains, event-types for vision might be more significant than an event-type for smell, but this might not be the case for dog brains. Some networks will be at the same level because their Groups have the same priority. Each Group network consists of all the associations of that Group connected together in a way (Figure 4) that shows which associations followed which in the past.

Only one node in each Group network can be active in each step and the active node, if there is one, will be the node for which all of its context events match the events in Short Term Memory during that step. PP's step-by-step experience is this collection of active nodes in the Group networks. Since each node has associated events predicting the next action or stimulus of a particular event-type, we can see that Long Term Memory is a predictor of what PP is about to do in the form of actions, and of what PP is about to experience in the form of stimuli from PP's body and world.

Now Long Term Memory really becomes alive because PP can "wander through" Long Term Memory planning, thinking, imagining, and looking for answers to questions. And it can do all of that inside its brain! We will see how that is done when we come to a discussion of planning.

If Long Term Memory could be made visible and if each of the active associations gave out a little light while it was active, we would be able to see the little lights going ON and OFF. When PP was 'confident' of what it was doing, the little lights would shine all the way up to the higher level networks, where the associations have a stronger grip on reality (more events in their contexts). When struggling to find its way or exploring, the little lights would shine mainly from the lower level networks.

PP has a simple mechanism for forgetting the least used associations when Long Term Memory is full, but it wasn't needed in the short interactions described in chapters 4 and 5.

Parallel Processing for Speed

Ceteris paribus, *slow nervous systems become dinner for faster nervous systems. Even if the computational strategies used by the brain should turn out not to be elegant or beautiful but to have a sort of evolutionary do-it-yourself quality, they are demonstrably fast.* (Churchland & Sejnowski, 1992, 2017)

PP is fast! Modules are fast. Each Group is designed to have its own processor, which grabs a context from Short Term Memory if it can. The context tells the Group processor where in Long Term Memory to find its associated events. The associated events from all the highest priority Group processors are thrown into a best-wins computation to decide on the robot's actions. Then the Group processors wait for the selected actions to be implemented by the body and world. Planning is even faster because it doesn't have to wait for implementation by the body and world. LeakBack, described in the next section, is a slower process that carries on in the background. Implementing PP and simulating its body and world on a serial computer is not fast.

There are about 100 billion neurons in the human brain, each of which operates in parallel (at the same time) at a time scale around a thousandth of a second. A modern personal computer has processors that operate serially (one step after another) at a time scale around one tenth of a billionth of a second. Taking into account the complexity of basic operations the modern personal computer is capable of, it is probably not far behind the brain in computing speed. This was certainly not the case when PP was invented more than 40 years ago.

There are advantages in parallel processing, apart from speed. Separate modules for processing raw sensory information from the eyes, from the ears, from touch areas, from taste buds, and from smell can be designed and tested separately and may even be able to self-heal.

Each Group of PP is designed to be a separate processor. In a robot body with sensors and effectors comparable to the human body, there could be several hundred or more Groups.

In the two main interactions described in chapters 4 and 5, BodyMoves, Speech, and Hand-Pointing are all performed sequentially, one after the other. All the stimuli from the World, however, are processed at the same time. There is no reason why the actions shouldn't be performed at the same time, instead of sequentially, but it is easier to follow them sequentially. The PP program allows actions and stimuli to be processed in parallel by putting them in different streams.

Planning is another process which could be carried out in parallel with the main behaviour of the robot, but sequential processing was easier to understand. The possibility of planning being done in parallel with the robot's main behaviour will become important when we discuss consciousness.

Parallel processing would be particularly appropriate for the LeakBack process.

LeakBack of Expectation

LeakBack is the process of calculating,[38] for each associated action event of a node of the network of a Group, the expectation of reaching a goal by performing that associated action event. An expectation is also calculated for each node from the expectations of its associated events. This is done after each action step of the PP brain. It applies only to Groups with associated event-types that are actions. Groups with associated event-types that are stimuli do not have networks and LeakBack, for the reasons given above.

The LeakBack process starts from the goals, marked on associated events of the network, and works backwards to the nodes possessing those associated events, and then back to the associated events pointing to those nodes, and back to the nodes possessing those associated events, and back to …

Because of the large amount of calculation involved, when carried out by a serial computer, LeakBack in the PP brain is limited to a maximum depth and minimum magnitude of expectation. In a biological version of the PP brain, LeakBack might be implemented by the continuous leaking back of a chemical, such as nitric oxide, through the physical network of neurons.[39]

The calculated expectations enable the PP brain to choose its best actions for each step.[40] (LeakBack is a different process from error back-propagation in neural networks.)

Planning Ahead

It is not too improbable, we feel, that consciousness is in some essential way the capacity to make one's own Plans and that volition is the capacity to execute them. (Miller, Galanter, & Pribram, 1960)

The actions taken by the PP brain to move the effectors (muscles) in its robot body are all carried out in the **Here-and-Now**. Planning is carried out in the **There-and-Then** (not Here-and-Now!).

When the PP brain makes a plan, it moves in the There-and-Then through its Long Term Memory to find a path ahead until it reaches a goal. When it finds such a path, it tries to follow the path to the goal in the Here-and-Now.

For planning, the PP brain needs another Short Term Memory to hold all the events for the associations it is using, so a copy of the Here-and-Now Short Term Memory is made before the making of a plan starts. The copy is called the STM-plan and it changes as the PP brain makes its plan.

The big difference between ordinary behaviour in the Here-and-Now and planning in the There-and-Then is that during planning the body and world are not providing stimuli and the actions are not being implemented by the body. While PP is in the Here-and-Now, the associations with stimuli as associated events are being updated all the time, but, when the PP brain is planning, they are used to predict stimulus events.

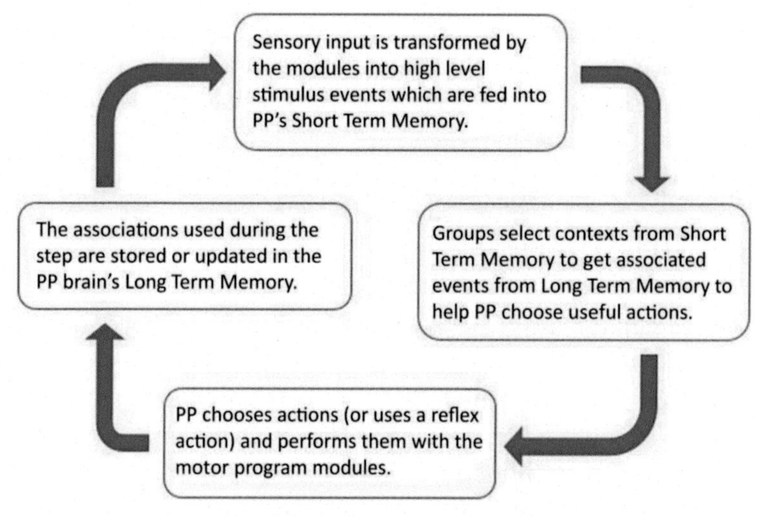

Figure 5 Basic Cycle.

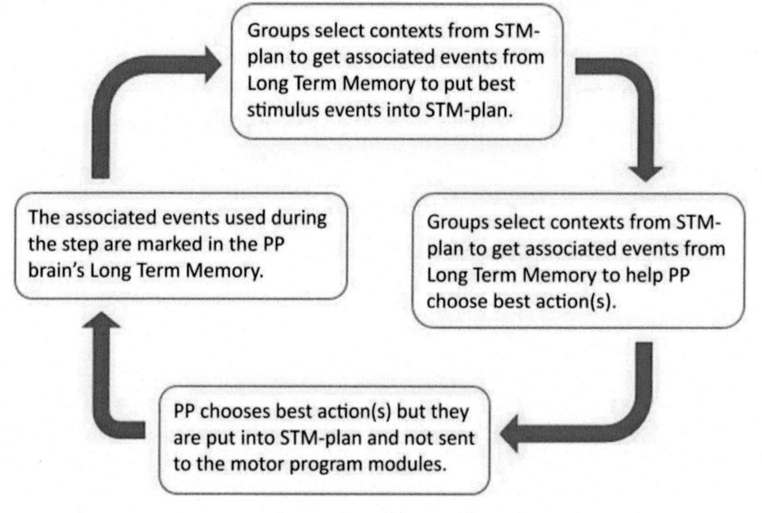

Figure 6 Plan Cycle.

When we come to discuss consciousness, we will find that this alternation of the basic cycle with planning avoids an important issue. In the meantime, it simplifies the behaviour by keeping the Here-and-Now separate from the There-and-Then.

How PP Works

PP interacts with the world in a series of steps. Each step follows the basic cycle of Figure 5. In the particular robot body and World used for the interactions described in chapters 3–5, this cycle is performed once in each **substep.** Several substeps make a **step.** The step begins with a substep for the movement of the robot body. This is followed by several substeps for speaking and pointing with the robot's Hand. The step ends when the speaking and pointing end.

More information about the PP program is given in Appendix-1.

At the end of each step, the PP brain tries to make a plan. The cycle used in the step and substeps of the plan is shown in Figure 6.

If the plan is successful in reaching a goal, PP starts the next step trying to follow the plan, which has been marked in its Long Term Memory. If the plan fails to reach a goal and is abandoned, the next step is just a repeat of the basic cycle.

Plan-following also follows the basic cycle of Figure 5 with the exception that when one of the marked associated events in Long Term Memory is used, its expectation is doubled to increase the probability that the plan will be followed.[41]

Chapter 3

Interacting with PP

Word learning really is a hard problem, but children do not solve it through a dedicated mental mechanism. Instead, words are learned through abilities that exist for other purposes. These include an ability to infer the intentions of others, an ability to acquire concepts, an appreciation of syntactic structure, and certain general learning and memory abilities. (Bloom, 2000)

People use language for doing things with each other, and their use of language is a joint action. (Clark H. H., 1996)

In order to demonstrate that PP can learn language in as short an interaction as possible, I have to put PP into a simple, but versatile, robot in a simple, but task-rich, world and choose conditions that others will see as a reasonable representation of how an infant learns. An infant learns language from other humans, so PP in its robot learns from another identical robot, controlled by me.

Language consists of words, so PP learns words instead of going through a babbling stage. Infants have to learn the relationship between hearing sounds and making sounds, and later between hearing words and saying words. PP hears the words it says and uses imitation, justified by the existence of mirror neurons, to help the learning of this relationship. The result is an interaction of only a thousand steps, with PP doing what it is told to do, saying what it has done, answering questions, and learning to take turns.

PP is going to be taught a number of tasks in a little World with two robots, as shown in Figure 7. The PP brain is the brain for the **blue robot** facing East and, as teacher, I am the brain for the **red robot** facing West. This means that PP is learning from a robot just like it. My **red robot** is called robot A, with A for adult, while PP's **blue robot** is called robot B, with B for Baby.

The World is a board with 4 horizontal rows of 5 squares each. In addition to the squares the robots are standing on, two of the squares have immovable Stumps, a **brown Stump** on the left and a **black Stump** on

An AGI Brain for a Robot. https://doi.org/10.1016/B978-0-323-85254-8.00003-X
© 2021 Elsevier Inc. All rights reserved.

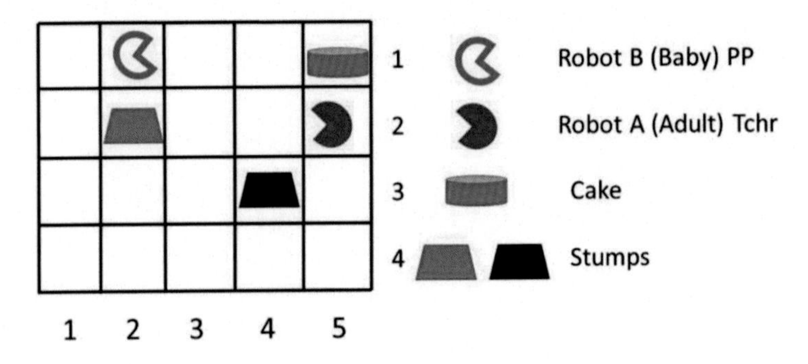

Figure 7 The World.

the right. Also there is a **green Cake** that can be pushed around and is the centre of attention. The board is surrounded by a Wall, which is not shown.

To consume the Cake, a robot must Squash the Cake and the Cake cannot be Squashed unless the other robot helps. My task is to teach PP to take turns with robot A in Squashing the Cake. PP has to learn how to Squash the Cake and has to learn to take turns. To do this, it has to use Working Memory.

One interaction between the two robots is described in chapter 4. A second interaction is described in chapter 5. Excerpts from these interactions are given in Appendix-4. The descriptions in chapters 4 and 5 explain the main features of the interactions with the minimum amount of detail. The arguments of this book are based on the evidence provided by these interactions together with historical evidence summarized in chapter *6*.

After what for me was a very long first interaction and robot A had been the last to Squash the Cake 194 steps earlier, PP moved forwards and said:

Next Turn Is Mine So You Wait

on step 995.[42] I made robot A WAIT and say

Good, Next Turn Is Yours So You Squash Cake.

PP then Squashed the Cake and said

I Squashed Cake So Next Turn Is <Yours.

PP had used Working Memory to remember whose turn it was to Squash the Cake.

I was very relieved, not only because PP had done the right thing but also because I could look into PP's Long Term Memory and use the encodingism detector to check what it had really learned. This was not just a lucky chance, as we shall see later.

More about the World—Actions

Both robots have BodyMove actions FORWARD, LEFT, RIGHT, WAIT, PAT, and SLAP. BodyMove is the event-type and the actions are events. FORWARD moves a robot one square forward on the board, if it can. LEFT and RIGHT are turns which rotate a robot 90 degrees to the left and right in the square that it is occupying. WAIT does nothing and is very useful. If PAT is done within reach of the other robot, it is approving to the other robot. Similarly, SLAP, done within reach of the other robot, is disapproving (punishing, if you like). Of course, I, the teacher in robot A, can disregard PATs and SLAPs from robot B.

If a robot is up against the Wall, a Stump, or the other robot, a FORWARD action has no effect, except that the robot is automatically disapproved (punished). If the Cake is in front of the robot when it does a FORWARD action, the Cake is Pushed, Squeezed, or Squashed. If there is an empty square beyond the Cake or just a single empty square to one side of the Cake, the Cake is **Pushed** into the empty square. If there are two empty squares, one on each side of the Cake, but it is blocked beyond, then the Cake is **Squeezed** into one of the two empty squares, randomly chosen. If there is no empty square into which the Cake can be Pushed or Squeezed, it is **Squashed.** Squashing causes the Cake to vanish (be consumed by the robot doing the Squashing) and a new Cake to appear at random in one of the other empty squares of the World. The robot that does the Squashing is automatically approved.

Each robot also has a Hand-Pointing action, of event-type **Point**, to enable it to point at the other robot, itself, the Cake, the Stumps, or the corners of the board.

A robot's Speech action, of event-type **InSSay**, comprises a Word with Intonation and Stress. The Intonation part of a Speech action is FLT (Flat), RIS (RISing for questions), CMD (for commands), DRP (Drop), YES (for approval), or NEG (for disapproval). These Intonations control the conversation between the robots. The DRP and CMD Intonations indicate that a robot is finishing both speaking and pointing. A Word is just a name for a sound that the robot hears. By convention, its first letter is in upper case. A Word can be Stressed, but usually isn't, because Stress is used in an important way. Both Intonation and Stress are used as **auxiliary actions,** Intonation to control the conversation and Stress to support Working Memory. An auxiliary action is an action which is not part of the specific task in hand.

I avoid all the problems associated with segmenting a sound sequence into Words by having the input to the robot already segmented.

The Words help teacher to teach consistently. By the end of the interaction, some of the Words are beginning to have some significance for PP. Teacher can introduce a new Word, with Intonation and Stress, at any time. From the start of the interaction, three Words are used by the robots with special meanings for the organizing PP program:

> ***Q***, short for DRP-*Q*, has a DRP Intonation and is treated by the PP program as meaning that the robot has finished speaking or is unable to speak, so it's the other robot's turn to do a BodyMove.
>
> ***CQ***, short for CMD-*Q*, has a CMD Intonation and is also treated by the PP program as meaning that the robot has finished speaking. Teacher uses it for ending commands.
>
> ***Q?***, short for RIS-*Q*, has the RISing Intonation and is treated as if the robot had asked a question, so the PP program switches the other robot to Speech and Hand-Pointing without an intervening BodyMove.

These are silent Words (***Q*** for Quiet) for ending speaking and for enabling the robots to take turns in speaking. Also, they enable PP to be taught the difference between statements, commands, and questions.

Two other Words, with built-in significance, enable the robots to approve (reward) and disapprove (punish) each other:

Good, short for YES-*Good,* has an Intonation YES which the receiving robot treats as approval of its Speech and Hand-Pointing.

Bad, short for NEG-*Bad,* has an Intonation NEG which the receiving robot treats as disapproval of its Speech and Hand-Pointing.

Stimulus Events

Each robot has the same repertoire of stimulus event-types. The robot's body has eyes for vision, touch sensors for feeling, ears for hearing, and a Hand for pointing. Eyes can't see behind the robot, and touch sensors can't feel behind the robot. These sensors provide the robot with the following sensory information:

Where the other robot is and which way it is facing, if visible.

What BodyMove the other robot has just done, if visible.

Where the Cake is, if visible.

What it is touching in front and to left and right.

Where the Stumps are if visible and which Stump is being touched.

Who is pointing at what, in so far as they can be seen.[43]

What it and the other robot are saying.

Imitation information, to be discussed below.

```
(The algorithm is somewhat more complicated when some Body
Moves are blocked.)

If robot B is facing the Wall:
    with 85% probability turn RIGHT, otherwise turn LEFT.
If robot B is facing a Stump or the other robot:
    with 85% probability turn LEFT, otherwise turn RIGHT.
If robot B is facing the Cake or a Space:
    with 85% probability move FORWARD;
    with 5% probability turn LEFT;
    with 5% probability turn RIGHT;
    otherwise WAIT.
```

Figure 8 The Reflex Motion Algorithm.

Reflex Motion

Because the PP brain starts with a blank slate (no associations) it can't do anything until it has learned some actions to select. To enable it to move around the World when it doesn't know what to do, the PP program gives each robot **Reflex Motion**, as described by Figure 8, which takes over when the PP brain is unable to decide on a BodyMove. Reflex Motion is random with a bias that makes the robot tend to move around the board. Of course, teacher doesn't need to let robot A be controlled by Reflex Motion.

Learning by Imitation

With knowledge of these [mirror] neurons, you have the basis for understanding a host of very enigmatic aspects of the human mind: "mind reading" empathy, imitation learning, and even the evolution of language.

(Ramachandran, 2000, 2011)

Since the discovery of **mirror neurons,**[44] it has become clear that humans, and some animals, have innate mechanisms for detecting how others are behaving. This is my justification for using an innate imitation mechanism in PP. If PP sees the other robot do a BodyMove in a certain context, PP stores the BodyMove and context in associations so that if it encounters the same context it can perform the same BodyMove. This kind of imitation is also done with Hand-Pointing but in this case PP must also be able to see what the other robot is pointing at. Of course, mirror neurons are not needed for imitating speech, since speech can be heard, but heard words still have to be converted to the saying of words. Presumably a human infant learns this conversion by experimenting with its voice in babbling. PP is assumed to have this conversion done by imitation, so we can avoid the babbling stage.[45] Hearing is imitated by saying. I can then treat imitation similarly for Body Moves, Speech, and Hand-Pointing.

The stimulus event-types include some that detect what the other robot is doing, touching, and seeing, when these activities of the other robot are visible. Robot B sees what robot A is doing as if it were doing it itself. This makes imitation easy when the other robot and its context are visible.

Learning also takes place by exploration with or without reinforcement, and by generalization using weaker contexts.

```
    BodyMoves            Speech             Hand-Pointing       Priority

    BA     BB         SA       SB            PA     PB          Highest
       BC             SC       SD         PC     PD     PE
BD     BE     BF   SE      SF       SG       PF     PG
BG     BH     BI   SH      SI       SJ       PH     PI
                      SK       SL                PJ             Lowest
```

Figure 9 Hierarchies of Action-Predicting Groups.

Groups for the Robot Body and World

The PP brain has a number of Groups for each of the action event-types. Details are again put into figures. Figure 9 shows the hierarchies of 31 action-predicting Groups. Each of these Groups has a network. The 2-letter Group names give the associated event-type with the first letter: B for BodyMove, S for Speech action, and P for Hand-Pointing. The second letter distinguishes the Group and indicates its priority, earlier letters in the alphabet being of equal or higher priority.

There is one other Speech-predicting Group, SW, provided for Working Memory. When associations of Group SW are used, only the most recent associated event (the one at the head of the list) is considered. In this way, the last association with a particular context is remembered for as long as it is not superseded.

Details of the Speech-predicting Groups, SA to SW, are given in Figure 10. SA and SB are the largest and highest priority Groups. They can hold Speech sequences of length 8 and 7 respectively.

The Working Memory Group SW has 3 HearSelfOrOther event-types delayed by 1 step, 2 steps, and 3 steps. HearSelfOrOther is the event-type "Hear own speech if speaking, else hear other robot's speech." The other HearSelfOrOther event-type in Figure 10 has an asterisk to indicate that the Word is stored without Intonation or Stress.

Group:	SA	SB	SC	SD	SE	SF	SG	SH	SI	SJ	SK	SL	SW
Event-type						**Delays**							
BodyMove	1	1	-	1	-	-	-	1	-	0	0	-	-
OtherRobotMove	1	-	-	-	-	-	-	-	-	-	-	-	-
OtherRobotDirn	-	1	1	-	1	1	1	-	1	-	-	-	-
OtherRobotPosn	1	1	1	1	1	1	1	-	1	-	-	-	-
ObjectShape	-	1	1	-	-	-	-	-	-	-	-	-	-
ObjectPosn	1	1,2	1	1	1	1	1	-	1	-	-	-	-
BrownStumpPosn	1	-	1	-	1	1	1	-	1	-	-	-	-
BlackStumpPosn	1	-	1	-	1	1	1	-	1	-	-	-	-
TouchFront	-	2	-	-	-	-	-	-	-	-	2	-	-
Touch	1,2	1	-	1	-	-	-	1-3	-	1,2	1	-	-
PointSelfOrOther	1	1,2	1,2	1	1	1	1	-	-	1,2	-	1,2	-
HearSelfOrOther	2-8	2-7	2-5	1-4	2-4	1,3,4	1,2,4	1-3	-	2	-	1-5	1-3
HearOther	1	1	1	-	-	-	-	-	1	1	-	-	-
IntonStressSay	1	1	1	-	-	-	-	-	2	2	-	-	-
HearSelfOrOther*	-	-	-	-	-	-	-	-	-	-	1	-	-
Twin Groups:				SOD					SOI		SOK		

* means Word is stored without Intonation or Stress.

Figure 10 Event-Types and Delays of Speech Action Groups.

The Groups SD, SI, and SK have twin Groups for **imitation** SOD, SOI, and SOK, respectively. The Groups for SOD, SOI, and SOK can be obtained from those for SD, SI, and SK by changing event-types from those of this robot to the corresponding ones referring to the other robot, such as BodyMove to OtherRobotMove. Also, the delay of HearSelfOrOther must be changed from 1 to 2, and the delay of InSSay from 2 to 1. (InSSay is Speech including Intonation and Stress.) These changes are such that going from SD to SOD, or SI to SOI, or SK to SOK transforms the contexts of associations from those of the viewing robot to those of the other robot, as required by imitation.

There are fewer Groups for BodyMoves and Hand-Pointing, but they are similar to the Groups for Speech.

The choice of the 31 action-predicting Groups in Figure 9 was guided by the principles listed in Figure 11.

The 35 stimulus-predicting Groups (not listed) included 6 for imitation and the remainder for planning. I was prevented from being more generous with the provision of Groups by the computing time taken by my serial computer.

The underlying computational power of the PP system arises from the majority view of predictions from active nodes (collections of associations with the same current context) of Groups at the same (highest) level.

(1) Groups should be designed for the robot Body and World, not for specific tasks.
(2) A set of Groups should prescribe a wide range of different contexts spread over the stimulus and action spaces.
(3) More specific Groups should be placed higher in the priority hierarchy and more general Groups should be placed lower. In particular, if the context event-types of one Group are a proper subset of the context event-types of another Group for the same associated event-type, the one must be placed below the other in their priority hierarchy
(4) If Groups are at the same level in a hierarchy, their context event-types should share as little information as possible.
(5) Imitation Groups should usually have a low or medium priority so that they apply to a wide range of situations.
(6) Sequences of words in contexts should not be too long. An early experiment with nursery rhymes confirmed that George Miller's famous magic number 7 ± 2 was about right.

Figure 11 Choosing Groups.

The majority view process has several features. When nodes are used for selecting an action (BodyMoves, Speech actions, or Hand-Pointing), nodes from Groups higher in the hierarchy get preference; that is if any predictions are made by the nodes of one Group, predictions from Groups below it in the hierarchy are disregarded (but they are still learned with the selected action!). If several predictions are made by one or more nodes of equal priority in a hierarchy, then each prediction is weighted by expectation values computed by LeakBack. The prediction with highest overall weight is chosen. Where a prediction from the Working Memory Group SW corroborates the prediction of Speech from an ordinary Group, the latter prediction is given additional weight.

Learning begins with PP using nodes of the more general imitation Groups and then moving up the Group hierarchy to stronger, more specific Groups to make the required discriminations. If a task is already supported by nodes of strong Groups, it will not be affected by changes to the more general associations resulting from a new task. For this reason, if PP learns a task well, with sufficient repetition, it will be less likely to be confused later by learning new tasks.

Aside-3. A mistake in my programming of a Hand-Pointing event-type prevented robot B from seeing what robot A was pointing at in the first interaction. My description of both interactions will omit the Hand-Pointing to make things simpler to follow, but the error was corrected in the second interaction, so the excerpt from the second interaction in Appendix-4 includes the Hand-Pointing.

The First Interaction

In the first interaction (described in chapter 4), teacher (me), acting through robot A, is trying to teach robot B. It is not all plain sailing. If robot B does something right or wrong and robot A can't see it or reach it, then teacher can't make robot A **PAT** or **SLAP** robot B. Also, there are times when teacher needs to do an action and has to choose between doing that action or **PATting** or **SLAPping.** Another problem for teacher is that the robots need to be going in opposite directions around the World so as to Squash the Cake in the corners, but to imitate robot A, robot B must be in the same context as robot A was. Teacher must ensure that the robots swap directions now and again.

The simple World shown in Figure 7 can be represented graphically with text:

```
-E--C
-#--w
---#-
-----
```

It doesn't take long to get used to the compact graphic and it is essential if you want to follow the excerpts from the interaction in Appendix-4.

Now blue robot B, which has the PP brain, is represented by a capital E because it is facing East (North is upwards). It is easy to remember that PP is uPPer case! Red robot A, with teacher controlling it, is a small red w because it is facing West. The Cake is a green capital C. The Stumps are brown # on the left and black **#** on the right. All the other squares are empty: "-".

The first step of the interaction is as follows, omitting the Hand-Pointing. Robot B does a **FORWARD** move **(B:F)** using Reflex Motion and says nothing (*Q*). Then robot A turns **RIGHT (A:R),** says

Next Turn Is Yours So You Move Forward,

and ends with a silent commanding Intonation (*CQ*). This, followed by the graphic representation, can be shown compactly as:

```
1
 B:FORWARD Q
 A:RIGHT Next Turn Is Yours So You Move Forward CQ

                         -E--C   1  --E-C   1  --E-C
                         -#--w  B:F -#--w  A:R -#--n
                         ---#-      ---#-      ---#-
                         -----      -----      -----
```

This is how the first step of the first interaction is shown in Appendix-4. I will be simplifying things further in the discussion to follow. In the graphical representation of the movements in the World, FORWARD is abbreviated to F, RIGHT is abbreviated to R, LEFT to L, WAIT to W, PAT to P, and SLAP to S. After robot B does a FORWARD move using Reflex Motion (B:F), it can be seen that the **blue E** has moved one square to the right (i.e. to the East).

CQ was explained above as meaning that teacher intended the Words from robot A to be a command *(You Move Forward).*

The '1' at the beginning of the first step is for '**step** 1'. A **substep** is either a BodyMove or Speech (plus Hand-Pointing not shown here), so the first step contains 12 substeps: 2 BodyMoves, 8 Words, and 2 silent terminators, *Q* and *CQ.*

With so much happening in a single step, it is difficult to give the reader the feeling of being robot A. The whole of this interaction is 997 steps and 11,540 substeps, so we will be taking a brief look at only a few places where significant learning has happened.

I will continue to omit the Hand-Pointing in the following description. The final excerpt in Appendix-4 does have the Hand-Pointing for the second interaction. The excerpts are quite readable with the help of chapters 4 and 5.

If you allow 5 seconds for each step, then the first interaction amounts to less than an hour and a half of learning in an infant's life, but it is concentrated learning. I did everything I could to stop PP from wandering off and doing its own thing, because I was trying to show that it is capable of using Working Memory with language. Teaching PP to remember numbers and to say what it was touching came towards the end of the interaction. By the end of the interaction, PP had stored 142,550 nodes in Long Term Memory, of which 62,911 were Speech action and stimulus nodes.

Chapter 4

Learning to Take Turns

In this first interaction, robot B, with the PP brain, is being taught

(i) to do the BodyMoves which robot A tells it to do;

(ii) to say what BodyMove it has just done;

(iii) to learn to write to, and read from Working Memory;

(iv) to use its Working Memory to take turns with robot A in Squashing the Cake[46];

(v) to turn away from the Wall after robot A Squashes the Cake;

(vi) to make a number and remember the last number it made; and

(vii) to talk about touching.

The PP program, data, and both interactions are available on the internet in ResearchGate.

It is important that my teaching through robot A is consistent because inconsistency would confuse PP and make the interaction even longer.

Aside-4. Nonsense! It wouldn't confuse PP at all, but would enable PP to do all sorts of things that teacher didn't want it to do.

On a first reading the reader might like to skim over this and the next chapter.

The interaction[47] is started with the robots in the positions shown in Figure 7. This is chosen to give a high probability that robot B will Squash the Cake in the first few moves. When PP can't choose a BodyMove action, Reflex Motion chooses an action using the algorithm in Figure 8. The first 5 BodyMoves of robot B are all by Reflex Motion, and are FORWARD, FORWARD, FORWARD, RIGHT, and FORWARD. I, as teacher, make robot A do the following actions:

An AGI Brain for a Robot. https://doi.org/10.1016/B978-0-323-85254-8.00004-1
© 2021 Elsevier Inc. All rights reserved.

1. B moves FORWARD, *Q* (says nothing)
   A turns RIGHT, says *Next Turn Is Yours So You Move Forward CQ*
2. B moves FORWARD, *Q*
   A WAITs, says *Next Turn Is Yours So You Squash Cake CQ*
3. B moves FORWARD and Cake is Squashed, *Q*
   A turns LEFT, says *You Squashed Cake So Next Turn Is <Mine Q*
4. B turns RIGHT, *Q*
   A moves FORWARD, says *I Moved Forward Q*
5. B moves FORWARD, *Q*
   A moves FORWARD, says *I Moved Forward Q*
6. **PP moves B FORWARD,** ***Q***

So, on step 6, PP does its first BodyMove! Remember, ***Q*** is for a default Quiet ending and ***CQ*** is for a command Intonation ending. (The Stress symbol **<** will be explained later.) Starting with the position in Figure 7, I can show the 5 positions after each BodyMove of robot A with these graphical configurations of the World:

```
-E--C   1  --E-C   2  ---EC   3  ----E   4  ----S   5  -----
-#--w  B:F -#--n  B:F -#--n  B:F -#--w  B:R -#-w-  B:F -#w-S
---#-  A:R ---#-  A:W ---#-  A:L ---#-  A:F ---#-  A:F ---#-
-----      -----      -----      --C--      --C--      --C--
    F = FORWARD     L = LEFT       R = RIGHT          W = WAIT
```

Looking at the **blue E**, you can see robot B doing three FORWARD moves to the East, the third one Squashing the **Cake C** because robot A is below the Cake preventing it from being Pushed South. (When the Cake is Squashed it reappears on one of the other empty squares chosen at random. Look for it on the bottom row.) On the fourth move, robot B turns RIGHT (**B:R**) towards the South and on the fifth step, robot B moves South with a FORWARD (**B:F**). On the sixth step, instead of a Reflex move, PP makes its first BodyMove with a FORWARD (**PP:F**).

```
-----    6   -----
-#w-S  PP:F  -#w--
---#-        ---#S
--C--        --C--
```

PP does this **FORWARD** move using 2 nodes from the low priority Groups **BH** and **BI,** shown in Figure 9:

Group **BH:**

[BH: BodyMove(-1), TouchFront(-2), TouchAll(-1), HearWord*(-1) ▶ BodyMove(0)]

with the node **0BH** stored on step 2 after robot B's move **FORWARD:**

[0BH: FORWARD, TouchSpace, TouchWallSpaceSpace, *Q* ▶ FORWARD]

and Group **BI:**

[BI: TouchFront(-1), Point(-2), (-1), HearWord(-5),(-4),(-3),(-2),(-1) ▶ BodyMove(0)]

with the node **3BI** stored on step 5 after robot B's move **FORWARD:**

[3BI: TouchSpace, PointMe, PointMe, *Q, I, Moved, Forward, Q* ▶ FORWARD]

Node **0BH** makes sense because robot B's second **FORWARD** is done when it is touching the Wall on its left: The event-type TouchAll(-1) in **BH** has the event **TouchWallSpaceSpace** with Wall to the left and Space in front and to the right, in **0BH.** Also it is not hearing anything because the last Word said by robot A was ***CQ*** **and** the * indicates that HearWord* is stored as ***Q*** without Intonation or Stress. The same context occurs again after robot A has made its fifth move so that this node tells PP to move **FORWARD.**

In Group **BI,** Point(-2),(-1) is an abbreviation for Point(-2), Point(-1). HearWord(-5),(-4),(-3),(-2),(-1) works the same way.

Node **3BI** is more tricky until one notices that robot A said

I Moved Forward Q

twice, once on step 4 and once on step 5. After the first time this was said, ReflexMotion made robot B move **FORWARD** and PP stored node **3BI.** When robot A said it again, node **3BI** told PP to move **FORWARD.** With both nodes (**0BH** and **3BI**) telling PP to move **FORWARD,** PP did the move on step 6.

Going into this amount of detail is tedious and you have now seen how it is done, so I will avoid more of it unless absolutely necessary.

PP does its first **RIGHT** turn on step 15 and this is after robot A told it ***You Turn Right CQ.*** PP does its first **LEFT** turn on step 48 and this is also after A telling it ***You Turn Left CQ.***

PP does its second move FORWARD on step 11 when robot A says

You Push Cake CQ

and it does Push the Cake. This sequence can be seen in the first excerpt of Appendix-4.

These actions are so early in the piece that it is obvious PP can't understand what it is doing.

How Much has PP Learned?

By the end of the 997 step interaction, PP has made 415 FORWARD moves, 177 RIGHT turns, and 168 LEFT turns. The FORWARD moves are more complicated because they include Push, Squeeze, and Squash, so I will look briefly at the 177 RIGHT turns.[48]

PP responds to robot A's ***You Turn Right CQ*** with a RIGHT turn 100 times, with a LEFT turn (wrong!) 11 times, and with a FORWARD move (wrong!) 6 times. On 17 steps, PP fails to respond and Reflex Motion responds with a RIGHT turn, a LEFT turn, or a WAIT. PP responds to a ***You Turn Left CQ*** with a RIGHT turn 4 times and to a ***Now You Wait CQ*** with a RIGHT turn once. PP performs a RIGHT turn, when not told to but for other reasons, 72 times.

We can say that PP has learned to do a RIGHT turn after the other robot has said ***You Turn Right CQ,*** but it has nothing to do with the movement name RIGHT being the same Word as the ***Right*** in the command. It would have learned the same associations (nodes) whatever the movement name for a right turn had been. Also, it has nothing to do with our understanding of the phrase ***You Turn Right.*** Any Words could be substituted consistently for ***You Turn Right*** and the learning would have been the same. PP has learned to link a particular Speech sequence with an action that turns the robot to the right through 90 degrees. <u>That statement avoids any encodingism</u>.

The phrase ***You Turn Right CQ*** said by robot A was accompanied by many different environmental configurations, all of which could have contributed to the nodes that enable PP to respond with a RIGHT turn.

In fact, PP uses 123 different nodes, the one used most being **43BI,** which was used 18 times. On 67 occasions, PP uses 2 nodes to choose **RIGHT,** and on 8 occasions it uses 3 nodes. Looking at Figure 9 for Groups at the same priority level in the BodyMove hierarchy, we see that PP could use 3 nodes with **BD, BE,** and **BF,** or with **BG, BH,** and **BI,** but it won't be often that nodes of all three Groups match the current events in Short Term Memory.

The other reason for so many nodes being used is that if a node is used for choosing an action in one situation, all valid nodes are stored or updated for that situation regardless of priority level.

The appropriateness of the node **43BI** is easy to see. It is the 43rd node of Group **BI** and the event-types of the context of **BI** are:

What the robot was touching in front of it.	Space for 43BI
The last two Hand-Pointings.	Point at Me, Point at Me for 43BI
The last five Words.	*Good You Turn Right CQ* for 43BI

There are 8 nodes of Group **BI,** in addition to **43BI,** which support the **RIGHT** turn after the three Words ***You Turn Right,*** all with the two Hand-Pointings, but with different touching and different Words preceding the three Words and ***CQ***:

5BI	Touching Wall;	*CQ*	before	*You Turn Right CQ*
24BI	Touching Robot;	*CQ*	before	*You Turn Right CQ*
43BI	Touching Space;	*Good*	before	*You Turn Right CQ*
44BI	Touching Wall;	*Good*	before	*You Turn Right CQ*
65BI	Touching Brown Stump;	*Good*	before	*You Turn Right CQ*
73BI	Touching Black Stump;	*Bad*	before	*You Turn Right CQ*
110BI	Touching Black Stump;	*Good*	before	*You Turn Right CQ*
112BI	Touching Space;	*Q*	before	*You Turn Right CQ*
117BI	Touching Robot;	*Good*	before	*You Turn Right CQ*

These nodes show how a group of nodes allows PP to learn to respond to a context ***You Turn Right CQ*** when there isn't a Group to prescribe the context exactly. Since Groups are not chosen for specific tasks, this is often the case.

Squashing the Cake

The main task being taught in this interaction with PP is to take turns in Squashing the Cake. To do this, PP has to remember who Squashed the Cake last time and it does that using Working Memory, which is made possible with a small special Group called **SW**:

[SW: HearWord(-3),(-2),(-1), ► Say+Stress(0)]

The nodes of this Group are treated differently from those of other Groups because only the associated event at the head of the list of associated events for a node is used. Recalling that the most recent associated event is always put at the head of the list of associated events, you will see that for the nodes of Group **SW** only the most recent associated event is ever used. The Group **SW** has another special feature. For a node to be stored under **SW**, its associated event must be a Stressed Word. When a node stored under **SW** is used for prediction, the Stress is disregarded. Figure 12 shows an example, taken from steps 253 to 256 of the interaction. PP appears to be doing very well, but its learning is still very fragile and it makes plenty of mistakes further on. A Stressed Word is marked by **<** in front of it.

Step
253 PP turns LEFT and says *I Turned Left Q.*
A turns RIGHT and says *You Push Cake CQ.*
254 PP moves FORWARD, Pushing Cake, and says *Next Turn Is Yours So You Move Forward CQ.*
A moves FORWARD and says *Good Next Turn Is Mine So You Push Cake CQ.*
255 PP moves FORWARD, Pushing Cake, and says *Next Turn Is Yours So You Squash Cake CQ.*
A moves FORWARD, Squashing Cake, and says *Good I Squashed Cake So Next Turn Is <Yours Q.*
256 PP turns LEFT and says *You Squashed Cake So Next Turn Is <Mine Q.*
A PATs and says *Good Q.*

```
--n--  253 --e--  254 ---e-  255 ----e  256 ----e
-#---  PP:L -#---  PP:F -#--C  PP:F -#--N  PP:L -#--W
---#C  A:R ---#C  A:F ---#N  A:F C--#-  A:P C--#-
----E      ----N      -----      -----      -----
```

Figure 12 Four steps from the Interaction.

Notice in Figure 12 that when, on step 253, PP correctly says

I Turned Left,

robot A is facing North so it didn't see PP's turn and therefore couldn't say ***Good*** to approve. Robot A then turns RIGHT and can see what is going on so that it can approve PP's saying

Next Turn Is Yours So You Move Forward CQ

on step 254 by saying:

Good Next Turn Is Mine So You Push Cake CQ.

On step 255, PP Pushes the Cake up into the corner and robot A Squashes it. A new **Cake C** appears at the beginning of the 3rd row. The graphics show the Board positions after robot A's moves in order to tell us the conditions under which PP in robot B moved. It is easy to deduce what the Board positions were after PP's moves. It is not easy to work out why PP did what it did.

PP turns LEFT on step 253 using two nodes (**126BA** and **127BB**) that it learned on step 135 when it was in a different position but one that looked the same to PP because it was looking out from the Board and had the Cake on its left, so it couldn't see robot A or the Stumps:

```
-----  135
-#-e-  PP:L
---#-
SC---
```

For similar reasons, when PP says

I Turned Left Q

on step 253, it uses nodes that it first learned on steps 80 and 84.

On step 254 (Figure 12) PP Pushed the Cake and said

Next Turn Is Yours So You Move Forward CQ.

To do this it used 13 nodes which it had first learned on steps 49, 59, 109, 137, and 138, but they are all weak, low priority nodes. In fact, it puts together the sentence by using nodes that take it step-by-step through the sentence:

113SL	*Q You Push Cake CQ*	▶	*Next*
114SL	*You Push Cake CQ Next*	▶	*Turn*
115SL	*Push Cake CQ Next Turn*	▶	*Is*
100SL	*Cake CQ Next Turn Is*	▶	*Yours*
35SL	*CQ Next Turn Is Yours*	▶	*So*
36SL	*Next Turn Is Yours So*	▶	*You*
339SK	TouchBkCkWl *You*	▶	*Move*
200SJ	TouchBkCkWl *You Move*	▶	*Forward*
198SH	TouchBkCkWl *You Move Forward*	▶	*CQ*

TouchBkCkWl is an abbreviation for the robot touching the Black Stump on its left (Bk), the Cake in front of it (Ck), and the Wall to the right of it (Wl). Notice in Figure 12 that after PP says this sentence on step 254, robot A starts its response by a FORWARD move, as requested, and says the Word ***Good*** which approves what PP has just said. The SL nodes also require the other robot to be pointing at 'me' i.e. PP. The SH, SJ, and SK nodes also require PP to have just done a FORWARD.

After doing the action FORWARD (Figure 12, step 254) that Pushed the Cake and after saying each Word, PP stores (or updates if already stored) nodes for all Groups that match the events, whether it used them for deciding or not. The storing and updating is done after robot A's response, so it knows from robot A's ***Good*** that the actions were approved.

Approval and Disapproval

The terms "reward" and "punishment" are too severe for what goes on in the teaching of PP, so I use "approval" and "disapproval". In the interaction between robot A (teacher) and robot B (PP) being discussed there is a lot of approval and disapproval because teacher is trying to get PP to learn fast. The BodyMoves PAT and SLAP by robot A have the effect of approving and disapproving the last BodyMove made by robot B. When PP in robot B imitates robot A by PATting and SLAPping, teacher knows that at this stage PP doesn't do the actions with any knowledge of what the actions do, so teacher disregards them.

The main difficulty with using **PAT** or **SLAP** is that they interfere with the sequences of BodyMoves that teacher is trying to teach robot B. Frequently, teacher in robot A omits **PATs** and **SLAPs** because doing some other BodyMove is more important for what is being taught.

Approval and disapproval of Words are done by the Word ***Good*** with **YES** Intonation and the Word ***Bad*** with **NEG** Intonation. Teacher can give this approval and disapproval only when it is robot A's turn to speak, so it applies to <u>all that robot B has just said</u>. Thus, on step 13 of the interaction (Appendix-4), robot B turned **LEFT** using Reflex Motion and then said

*I Turned **Right** CQ.*

Robot A turned **LEFT** and said

***Bad** You Turn Right CQ.*

The ***Bad*** applied to all of robot B's (i.e. PP's) Words *I Turned Right CQ* in spite of the fact that it was only the Word ***Right*** and ***CQ*** that were wrong. This has the effect of marking as disapproved all the associated events that were activated for these Words, not just the highest priority ones. With all this disapproval spread over several Words, I needed some way of getting out of the mess.

One way to avoid the problem would have been to let the robots interrupt each other during speaking, but this would have made the interaction more complicated and, in my view, less realistic. My solution was intended to reflect the way a parent helps a child who is hesitating while trying to find the next word. If PP is choosing from a number of Words, as a result of nodes matching Short Term Memory, and one of the Words is **blocked** by a disapproval mark, teacher in robot A is given the option of **unblocking** the blocked Word, or of **prompting** with a different Word. Unblocking is used a lot.

I see unblocking as having robot A interpose a Word that robot B is trying to say. Prompting is rarely used because it is likely to introduce new context and stop PP from speaking. In the interaction, prompting is not used until step 369 when it is used to make PP say ***Wait*** instead of ***Squash***. As a result, PP stops after saying it, ending with a default ***Q,*** instead of the correct ***CQ,*** which it would have ended with.

Teacher is given the option to unblock a Word only if PP has no alternative unblocked Word to choose with as high a priority. Therefore unblocking is allowing PP to choose its best Word for that step in the knowledge that most Words blocked will not be the reason for the blocking.

Good and ***Bad*** invite imitation by PP. Teacher tolerates this in this short interaction. Learning correct usage of ***Good*** and ***Bad*** is seen as something for well into the future when the PP brain is in a real robot and has acquired much more of the meaning of what it is doing.

Aerial View of Interaction

To give the reader a feel for the way the interaction went without plodding through the steps, I will 'fly over it' while commenting on significant features.

Since robot A doesn't usually make mistakes, its teaching is easier to understand than that of robot B, which is struggling to imitate and learn. Figure 13 lists what teacher says through robot A in the first 200 steps. We omit the Hand-Pointing as it is not making an important contribution. The successes of PP in the first 200 steps, not including chance correct responses by Reflex Motion or incomplete responses by PP, are listed in Figure 14.

The item that stands out in Figure 14 is that PP says

I Moved Forward Q

after doing a FORWARD BodyMove 35 times, when it heard robot A saying that only 17 times (26 – 9 in Figure 13). However, closer inspection shows that 26 of the 35 times were when the FORWARD was done after robot A told it to do a FORWARD, so a strong context was set up with

You Move Forward CQ FORWARD

to help it choose the

I Moved Forward Q.

Remember, PP mustn't say

I Moved Forward Q

when its FORWARD move Pushes, Squeezes, or Squashes the Cake.

You Move Forward CQ	52 times	
You Turn Right CQ	25 times	
You Turn Left CQ	21 times	
You Push Cake CQ	13 times	
Now You Wait CQ	5 times	
I Moved Forward Q	26 times	9 not seen by robot B
I Turned Right Q	10 times	6 not seen by robot B
I Turned Left Q	10 times	6 not seen by robot B
I Pushed Cake Q	3 times	2 not seen by robot B
I Squeezed Cake Q	3 times	2 not seen by robot B
I Squashed Cake So Next Turn Is <Yours Q	5 times	
You Squashed Cake So Next Turn Is <Mine Q	7 times	
Next Turn Is Yours So You Move Forward CQ	3 times	
Next Turn Is Yours So You Squash Cake CQ	7 times	
Next Turn Is Yours So You Push Cake CQ	4 times	
Next Turn Is Mine So You Push Cake CQ	4 times	
Next Turn Is Mine So You Move Forward CQ	1 time	
Q	1 time	

Figure 13 What robot A says in the first 200 steps.

We needn't worry about the fact that PP hasn't said

I Turned Right Q

after turning RIGHT in the first 200 steps (Figure 14), because it has seen robot A do that only 4 times (10 – 6 in Figure 13). By the end of the 997 step run PP has said

I Turned Right Q

after turning RIGHT 111 times, of which 69 were after robot A told it to turn right.

Of greater interest is how PP is progressing with learning to Squash the Cake and talk about it. Teacher in robot A made sure that the Cake was Squashed only in the corners of the World and not between the Stumps and the Wall. There are 4 corners, 2 robots to do the Squashing and 2 ways for the robots to be positioned. This positioning is described as "clockwise AB" (abbreviated to **cAB**) and "clockwise BA" (abbreviated to **cBA**). If you look around the World in a clockwise direction, cAB will have you see A before B, and cBA will have you see B before A.

PP moved FORWARD when told to	44 times
PP turned RIGHT when told to	9 times
PP turned LEFT when told to	13 times
PP Pushed Cake when told to	11 times
PP WAITed when told to	5 times
PP says *I Moved Forward Q* after FORWARD	35 times
PP says *I Turned Left Q* after LEFT	9 times
PP says *I Turned RIGHT Q* after RIGHT	0 times
Next Turn Is Yours So You Move Forward CQ obeyed	2 times
Next Turn Is Yours So You Squash Cake CQ obeyed	6 times
Next Turn Is Yours So You Push Cake CQ obeyed	4 times
Next Turn Is Mine So You Push Cake CQ obeyed	4 times
Next Turn Is Mine So You Move Forward CQ obeyed	1 time
I Squashed Cake So Next Turn Is <Yours Q said	4 times
You Squashed Cake So Next Turn Is <Mine Q said	4 times
Next Turn Is Yours So You Move Forward CQ said	0 times
Next Turn Is Yours So You Squash Cake CQ said	1 time cAB SWcnr
Next Turn Is Yours So You Push Cake CQ said	0 times
Next Turn Is Mine So You Push Cake CQ said	0 times
Next Turn Is Mine So You Move Forward CQ said	0 times

Figure 14 The Successes of PP (robot B) in the first 200 steps.

Thus, in Figure 7, when robot B moves two squares to the right, it will be ready to Squash the Cake in a cBA orientation. So PP has to learn what to say in $4 \times 2 \times 2 = 16$ Squashing situations. The contexts in which these Squashes are carried out would have been halved if the colours and feels (Touch) of the Stumps could have been disregarded, because then there would have been no difference for the robots between Squashing in the NW and SE corners, or between the NE and SW corners.

The 4 times that PP says

I Squashed Cake So Next Turn Is <Yours Q

correctly in the first 200 steps were in the SW corner (cBA, step 104), in the SE corner (cAB, step 139), and in the NE corner (cAB, steps 165, 200) twice.

Robot		NW corner	NE corner	SE corner	SW corner
A	cBA	17,150	228,390	116,316,636	289,474,594
A	cAB	350,527	255,803	421	60,179,446,505
B	cBA	242,404	3,435,569	341,492,518	36,104
B	cAB	88,580	165,200,370,465,996	139	267,303,683

Figure 15 Steps when Cake or Roll is Squashed.

The 4 times that PP says

You Squashed Cake So Next Turn Is <Mine Q

correctly in the first 200 steps were in the SW corner (cAB, step 61), in the SE corner (cBA, step 117), in the NW corner (cBA, step 151), and again in the SW corner (cAB, step 180).

Figure 15 (and Appendix-2) shows the steps in the whole interaction where the Cake or **Roll** is Squashed. Between steps 500 and 600, the Cake is exchanged for a Roll to see how PP reacted to this change. On 5 different steps, PP calls the Roll a Cake without it upsetting the rest of its phrases. Then, after the Roll has been exchanged back for the Cake, it makes just one reference to Roll. Altogether, PP handles the appearance of a Roll in place of the Cake quite well.[49]

Step 61 is the first time that PP says

You Squashed Cake So Next Turn Is <Mine Q

correctly, and, apart from calling the Roll a Cake, it gets it right for the next 18 times that robot A Squashes the Cake/Roll to the end of the interaction. This isn't too surprising as robot A sets up the context for PP with

I Squashed Cake/Roll So Next Turn Is <Yours Q.

Step 88 is the first time that PP attempts

I Squashed Cake So Next Turn Is <Yours Q

and gets it wrong with <Mine for <Yours. From then on, it gets it right even when on step 580 the Squashing is done by Reflex Motion out of turn.

Remembering to Take Turns

PP uses the two learned sentences

You Squashed Cake So Next Turn Is <Mine Q

and

I Squashed Cake So Next Turn Is <Yours Q

to remember whether it is robot A's turn or PP's turn to Squash the Cake. The Stressed Word ***<Mine*** stores the association

Next Turn Is ▶ <Mine

in Long Term Memory with Group SW. The Stressed Word ***<Yours*** stores the association

Next Turn Is ▶ <Yours

in Long Term Memory with Group SW. Therefore, the most recently stored associated event with context ***Next Turn Is*** is either ***<Mine*** or ***<Yours,*** depending on whose turn it is next.

To read the memory, PP has to say the context ***Next Turn Is*** and then the associated event, either ***Mine*** or ***Yours,*** will be given to it (i.e. predicted) without the Stress by the SW association. Storing in Working Memory is done with Stress and then later in the interaction the stored Word can be recovered without Stress:

Next Turn Is <Mine <u>stores</u> *<Mine.* Later: *Next Turn Is* <u>reads</u> *Mine.*

Next Turn Is <Yours <u>stores</u> *<Yours.* Later: *Next Turn Is* <u>reads</u> *Yours.*

Robot B Squashes the Cake on step 3 using Reflex Motion. (Appendix-2 lists all the steps when the Cake or Roll is Squashed.) On step 17, robot A Squashes the Cake. On step 36, PP Squashes the Cake, still without saying anything. However, on step 60 when robot A Squashes the Cake in the SW corner and says

I Squashed Cake So Next Turn Is <Yours Q,

PP turns LEFT and says correctly

You Squashed Cake So Next Turn Is <Mine Q.

This is imitating what robot A said on step 3 after robot B had Squashed the Cake in the NE corner. The NE corner is very similar to the SW corner, but on step 3 the orientation of the robots was cBA, while on step 60 the orientation is cAB, enabling imitation.

PP's sentence is made by the very weak **SK** Group:

[SK: BodyMove(-1) TouchFront(-2) TouchAll(-1) HearWord*(-1) ► Speech(0)]

with associations:

[18SK: LEFT TouchFrontRobot TouchSpaceSpaceRobot *Q* ► *You*]
[19SK: LEFT TouchFrontRobot TouchSpaceSpaceRobot *You* ► *Squashed*]
[20SK: LEFT TouchFrontRobot TouchSpaceSpaceRobot *Squashed* ► *Cake*]
[21SK: LEFT TouchFrontRobot TouchSpaceSpaceRobot *Cake* ► *So*]
[22SK: LEFT TouchFrontRobot TouchSpaceSpaceRobot *So* ► *Next*]
[23SK: LEFT TouchFrontRobot TouchSpaceSpaceRobot *Next* ► *Turn*]
[24SK: LEFT TouchFrontRobot TouchSpaceSpaceRobot *Turn* ► *Is*]
[25SK: LEFT TouchFrontRobot TouchSpaceSpaceRobot *Is* ► *<Mine*]

On step 228 (Appendix-4) when robot A Squashes the **Cake** again, PP is using an array of stronger associations to say the same Words. These are stronger associations because they have more events in their contexts. Learning starts with weak associations and becomes established with stronger associations:

[168SE:N VOthPos42 ObPosU VBn/k:U/U PtMe HearFO:*Turn,Is,<Yours* ► *You*]
[172SF:N VOthPos42 ObPosU VBn/k:U/U PtYou HearFO:*Is,<Yours,You* ► *Squashed*]
[172SG:N VOthPos42 ObPosU VBn/k:U/U PtYou HearFO:*<Yours,You,Squashed* ► *Cake*]
[504SI: N VOthPos42 ObPosU HearOthr:Q InSSay:*Q, Cake* ► *So*]
[634SD:LEFT VOthPos42 ObPosU TouchSpWlRb PtMe HearFO:*You,Squashed,Cake,So* ► *Next*]
[180SC:VOthPos42 ObPosU VBn/k:U/U PtMe,Me HearFO:*You,Squashed,Cake,So* HearOthr:*Q* InSSay:*Next* ► *Turn*]
[635SD:LEFT VOthPos42 ObPosU TouchSpWlRb PtMe HearFO:*Squashed,Cake,So,Next* ► *Turn*]
[181SC:VOthPos42 ObPosU VBn/k:U/U PtMe,Me HearFO:*Squashed,Cake,So,Next* HearOthr:*Q* InSSay:*Turn* ► *Is*]
[636SD:LEFT VOthPos42 ObPosU TouchSpWlRb PtMe HearFO:*Cake,So,Next,Turn* ► *Is*]
[182SC:VOthPos42 ObPosU VBn/k:U/U PtMe,Me HearFO:*Cake,So,Next,Turn* HearOthr:*Q* InSSay:*Is* ► *<Mine*]
[637SD: LEFT VOthPos42 ObPosU TouchSpWlRb PtMe HearFO:*So,Next,Turn,Is* ► *<Mine*]

Disregard the details.[50]

On step 88, PP makes a mistake and says

*I Squashed Cake So Next Turn Is <**Mine** Q,*

which sets up Working Memory wrongly, so robot A lets PP Squash the Cake on step 104 to agree with Working Memory. For the next 28 Squashes to step 569, PP and robot A alternate Squashing correctly.

From step 500 to step 600, as mentioned before, teacher introduces a Roll in place of the Cake. This causes a little trouble with PP continuing to call the Roll a Cake, but the Squashing is done properly until step 580.

The situation on step 580 has robots A and B in the NW corner ready to Squash the Roll and robot A tells robot B

Next Turn Is Mine So Now You Wait CQ (R for Roll):

```
RW---
n#---
---#-
-----
```

PP has responded to A's

Now You Wait CQ

before but the situation is sufficiently different this time for it to fail to choose an action. Reflex Motion takes over and Squashes the Cake, when it was robot A's turn! PP sets up the Working Memory correctly with

I Squashed Cake So Next Turn Is <Yours Q.

The alternating taking of turns is upset, but the reading and writing of Working Memory is valid.

There is one more mistake before the last 4 Squashes go correctly. On step 603, out of the blue, PP says

I Squashed Cake So Next Turn Is <Yours Q,

when it hadn't Squashed the Cake, for which robot A said *Bad* in disapproval.

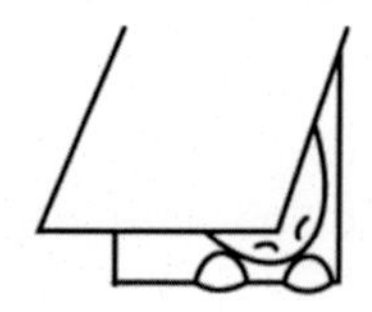

Reading Working Memory

Writing in Working Memory for taking turns by saying ***<Mine*** and ***<Yours*** is quite easy for PP because robot A sets up the context for PP, either by telling PP to Squash the Cake, or by telling PP that robot A has itself Squashed the Cake. Reading Working Memory isn't so easy because the context isn't usually set up for it by robot A. Also, Working Memory doesn't propose an action, but only strengthens one of the choices that PP already has. If PP has just one choice and it isn't the action that Working Memory is advocating, then teacher is given the chance to prompt with Working Memory's recommendation.

Up to step 164, PP reads Working Memory a few times and has a few prompts, but without completing the appropriate phrase. On step 179, robot A tells robot B

Next Turn Is Mine So You Push Cake CQ,

PP does so and replies

Next Turn Is Yours So You Squash Cake CQ.

The underlined ***Yours*** has to be prompted because PP has a strong association **286SB** stored on step 164 in a similar situation saying ***Mine.*** After this, in a similar situation PP should be able to choose the Word advocated by Working Memory because **286SB** will now be predicting the alternative choices ***Mine*** and ***Yours.*** And it does, on step 255 in the NE corner with ***Yours*** and on step 302 in the SW corner with ***Mine.*** On steps 369 and 995, **286SB** is supported by **583SA** to say ***Mine*** in the NE corner. In fact, this is how PP said on step 995

Next Turn Is Mine So You Wait Q,

which was quoted in chapter 3. (A table with all the Squashes is given in Appendix-2.)

Remembering Numbers

Between steps 606 and 716 of the interaction, teacher introduces PP to another task and gets PP in robot B to answer questions using the RISing Intonation ***Q?***.

This is how it should go:

Robot A says *Make Your Number Two CQ* (Similarly for One and Three.)

Robot B says *My Number Is <Two Q* (This puts Two in Working Memory.)

Later:

Robot A says *What Is Your Number Q?* (Question by RISing Intonation.)

Robot B says *Now My Number Is Two Q* (Read from Working Memory)

In order to show robot B what to do, robot A has to follow

Make Your Number Two CQ

with

My Number Is <Two Q

without robot B saying anything, so that robot B hears the second sentence immediately following the first. Then, when it hears the first sentence, robot B will have the context for saying the second sentence. Robot A has to disapprove several times when robot B speaks after the first sentence.

The Stress on the number (e.g. <Two) ensures that the number is put into Working Memory with the context

My Number Is.

If robot B kept moving so as to change the context in the World, the numbers would have to be learned in different contexts and the interaction would be longer. To keep the interaction short, robot B was guided by robot A to be in the same position as robot A had been in, so that robot B could imitate what robot A had demonstrated. When PP in robot B hears robot A say a Word and robot B can see what context robot A is in, PP stores what robot A said and the context in which robot A said it. Later, if PP finds itself in the same context, imitation enables PP to say the Word that robot A had said.

The second pair of sentences have their own difficulties. A different way of presenting the two sentences as consecutive is used: the two sentences are said as one separated only by **FLT-Q. FLT-Q** allows robot A to continue with the second sentence. Working Memory has to be read to find out what the last number is. Here, as with the Taking of Turns task, Working Memory can be read properly only if the ordinary associations are predicting the alternatives.

In order to make this happen in a short part of the interaction (110 steps), robot A gets robot B into the following configurations one after the other:

```
              -----              ----E              -----
        cAB:  -#---        cBA:  -#--e        cAB:  -#--C
              W--#-              ---#-              W--#-
              w-C--              -C---              w----
Steps:        605-620            643-673            687-717
```

Both robots are facing out from the board and the Cake and Stumps are not being seen. In the first configuration, robot B is on the right of robot A, in the second configuration robot B is to the left of robot A, and in the third configuration robot B is again on the right of robot A.

On steps 606 and 607, robot A says

Make Your Number Three CQ

and then, with PP **WAITing** and saying nothing, robot A says

My Number is <Three Q.

This ensures that when PP gets the command

Make Your Number Three CQ,

that command is the context for PP to respond with

My Number is <Three Q,

and the Stress on ***Three*** puts ***Three*** into Working Memory. This is repeated on steps 611 and 612 for the number ***<Two,*** and on steps 616 and 617 for the number ***<One.***

On step 609, robot A says

What Is Your Number FLT-Q Now My Number Is Three Q.

The sentence is given the ***FLT-Q*** in the middle so that it doesn't end the sentence there, which would have been the case if it had used ***RIS-Q*** or ***DRP-Q,*** the default sentence-ender. When PP hears the first part of the sentence with a ***RIS-Q*** to make it a question, it will answer with the second part <u>except</u> that the number will need to be read from Working Memory, using the context ***My Number Is,*** which was also used for storing the number in

My Number Is <Three Q.

This causes a certain amount of trouble because Working Memory only strengthens choices that PP already has; it doesn't suggest Words.

The process just described for the number ***Three*** on steps 606, 607, and 609 is repeated on steps 611, 612, and 614 for number ***Two*** and on steps 616, 617, and 619 for number ***One.***

On step 642, before robot A has turned towards the East and without being asked, PP says

My Number is <Three CQ.

Even if it had been asked, the ***CQ*** is wrong, so robot A says ***Bad.*** Needless to say, this bit of creativity is done using weak associations which disregard much of the context. Here are the nodes used:

[1343SK: WAIT TchF:WI WIWIRb HearSFO:*Q* ▶ *My*, et al]
[1348SK: WAIT TchF:WI WIWIRb HearSFO:*My* ▶ *Number*]
[1346SK: WAIT TchF:WI WIWIRb HearSFO:*Number* ▶ *Is, One, Two, Three*]
[3357SD: WAIT VOthPos42 ObPos:U WIWIRb PtMe HearFO:*CQ,My,Number,Is*
▶ *<Three, <Two, <One*]
[1347SK: WAIT TchF:WI WIWIRb HearSFO:Three >> CQ]

The one strong node, **3357SD,** gives an appropriate associated event ***Three*** for the wrong context ***My Number Is*** given by the previous three nodes. Again, disregard the details.

On step 645 (Appendix-4), PP barges in again, this time with

What Is Your Number Q?

and robot A answers

Now My Number Is Two Q,

which isn't actually correct because robot A set its number to ***<One*** on step 617. This mistake by teacher fortunately makes no difference. Robot A immediately (same step) asks robot B

What Is Your Number Q?

PP replies

Now My Number Is Three Q.

This is correct because on step 642 PP set its number to ***Three.***

Between steps 649 and 662 robot A demonstrated the sentences for the cBA configuration so that robot B could imitate it when they got back to a cAB configuration. Robot B had now been shown the sentences in both the cAB and cBA configurations. How would it do?

The ***Q*** ending to a sentence is a default ending and PP is not allowed to choose that ending for fear of it producing strings of ***Q's*** or getting into a limit cycle with it. However, PP must be allowed to choose ***CQ*** and ***Q?*** at the ends of sentences so that it can make a command or ask a question of robot A.

On step 663, robot A asks robot B

What Is Your Number Q?

and PP answers correctly

Now My Number Is Three Q.

Robot A says ***Good,*** but PP didn't use Working Memory properly to give the answer. ***Three*** was the only number being predicted and it agreed with Working Memory.

Two steps later, robot B is told

Make Your Number One CQ

and PP answers

My Number Is <One CQ.

The command ending ***CQ*** is wrong, but the reply is otherwise correct, so A says ***Good.*** On step 666, A asks

What Is Your Number Q?

and PP replies

Now My Number Is One^ CQ.

PP had only ***Three*** predicted and Working Memory had ***One,*** so robot A had to prompt **(^)** with the ***One.*** Teacher is now concerned that ***CQ*** has been said again and says ***Bad.***

To move this along, I summarize the rest of PP's responses to robot A's questions and commands (some can be seen in Appendix-4):

Step	PP's response	robot A	Comment
667	*My Number Is One Q*	*Bad*	ch:1,3 SW:1 Unasked no *Now*
669	*My Number Is <Two Q*	*Good*	
670	*Now My Number Is Two^ Q*	*Good*	ch:1,3 SW:2 Prompted
673	*My Number Is <Three Q*	*Good*	
	Change to cAB		
688	*Now My Number Is Three CQ*	*Good*	ch:1,2,3 SW:3 *CQ* wrong
689	*Now My Number Is Three CQ*	*Bad*	ch:1,2,3 SW:3 Unasked *CQ* wrong
693	*I Turned Left Q*	*Bad*	reply to *Make Your Number One CQ*
699	*My Number Is <Three Q*	*Bad*	asked to *Make Your Number Two CQ*
700	*Now My Number Is Three CQ*	*Good*	ch:3 SW:3 but *CQ* wrong
703	*My Number Is <One CQ*	*Bad*	asked to *Make Your Number Two CQ*
704	*Now My Number Is One^ Q*	*Good*	ch:3 SW:1 Prompted
707	*My Number Is <Two Q*	*Good*	
708	*Now My Number Is Two^ Q*	*Good*	ch:1,3 SW:2 Prompted
711	*My Number Is <One Q*	*Good*	
712	*Now My Number Is One Q*	*Good*	ch:1,2,3 SW:1
715	*My Number Is <Three Q*	*Good*	
716	*Now My Number Is Three CQ*	*Good*	ch:1,2,3 SW:3 but *CQ* wrong

The abbreviations need explanation:

ch:1,2,3 means PP is predicting *One, Two,* and *Three.*
SW:3 means Working Memory is saying *Three.*
If the choices predicted by PP don't include what Working Memory is saying, then teacher should prompt (^) with what Working Memory is saying.

The last entry (step 716) shows that

Now My Number Is Three CQ

is still stuck with the unwanted ***CQ,*** so teacher says ***Bad*** when PP repeats the sentence, unasked, on the next step, even though the number is correct.

It should be mentioned that part of the way through teaching this Number task, robot A Squashes the Cake in these three steps:

Step

635: A: FORWARD *Good Next Turn Is Mine So You Move Forward CQ*

636: PP: FORWARD *Bad Next Turn Is Yours So You Squash Cake CQ*

A: FORWARD (Squashes) *Good I Squashed Cake So Next Turn Is <Yours Q*

637: PP: RIGHT *You Squashed Cake So Next Turn Is <Mine Q*

PP hasn't forgotten that it is still taking turns with Squashing the Cake.

Talking About Touching

This task doesn't use Working Memory. The intention was to be more casual about the teaching, with robot A saying

Look I Am Touching Brown Stump Q

Look I Am Touching Black Stump Q

Look You Are Touching Brown Stump Q

Look You Are Touching Black Stump Q

What Am I Touching Q?

or

What Are You Touching Q?

and robot A or robot B touching the appropriate Brown or Black Stump in front of it. Robot A was also to be pointing at the Stump in question, but a program error in this interaction meant that robot B was not seeing what robot A was pointing at! More details are given in Appendix-3.

PP is not shown how to answer the questions, in the way PP was shown for remembering numbers, so it must use imitation of how robot A does it. In the 250 steps left before the end of the interaction, PP uses the first sentence correctly 6 times, the second 21 times, the fourth 6 times, and the sixth once. It hasn't managed the third and fifth.

For imitation to work easily, robot A needs to get the robots into pairs of positions with the two robots exchanging places, so that robot B is in the same situation as robot A was when it demonstrated a sentence. This was done for only 2 steps with the robots touching the Brown Stump, and for 15 steps with the robots touching the Black Stump. With more interaction PP would learn to use all the sentences when the correct Stump is being touched.

There is no suggestion that PP knows what it is doing. There is a small link between

Touching Brown Stump

and the actual touching of the Brown Stump, and between

Touching Black Stump

and the actual touching of the Black Stump. There is the beginning of a link between

What Am I Touching Q?

and

Look You Are Touching;

also a stronger link between

What Are You Touching Q?

and

Look I Am Touching.

The visual information available to PP about the Stumps is only their colours and their positions. So what could PP learn about touching in this simple World? Not much!

Here is a list:

How turning to left or right changes what is touched in front, on the left and on the right.

The shortest route to touch the Brown/Black Stump, robot A or the Cake/Roll.

Will a FORWARD move Push, Squeeze, or Squash the Cake/Roll it is touching in front?

Talking about the above.

Turning Away from the Wall

All but one of the tasks listed at the beginning of this chapter have been discussed. The exception is "to turn away from the Wall after robot A Squashes the Cake". This is how it went:

Step			success			corner		
18	B	LEFT	No	A SLAPs	cBA	NWcnr		Rand=16
61	PP	LEFT	Yes	A PATs	cAB	SWcnr	unblocked	12BI
117	PP	LEFT	No	A SLAPs	cBA	SEcnr		12BI
151	B	RIGHT	Yes	A PATs	cBA	NWcnr		Rand=0
180	PP	LEFT	Yes	A PATs	cAB	SWcnr		74BH
229	PP	LEFT	No	A SLAPs	cBA	NEcnr		71BI
256	PP	LEFT	Yes	A PATs	cAB	NEcnr	unblocked	148BG,74BH
290	B	RIGHT	Yes	A PATs	cBA	SWcnr		Rand=4
317	PP	RIGHT	Yes	A PATs	cBA	SEcnr		12BI
351	PP	LEFT	Yes	A PATs	cAB	NWcnr	unblocked	12BI
391	PP	RIGHT	Yes	A PATs	cBA	NEcnr		183BG,189BH
422	PP	LEFT	Yes	A PATs	cAB	SEcnr		71BI
447	PP	LEFT	Yes	A PATs	cAB	SWcnr		168BA,168BB
475	PP	LEFT	No	A SLAPs	cBA	SWcnr		71BI
506	PP	LEFT	Yes	A PATs	cAB	SWcnr		155BC
528	PP	LEFT	Yes	A PATs	cAB	NWcnr		12BI
595	PP	RIGHT	Yes	A PATs	cBA	SWcnr		236BC
637	PP	RIGHT	Yes	A PATs	cBA	SEcnr		239BD
804	PP	LEFT	Yes	A PATs	cAB	NEcnr		235BA

Looking down the 'success' column of 'No and Yes', we see that PP has done quite well: 3 No's, all before halfway, and 13 Yes's. However, all the No's were obtained using associations (column on the right) from the BI Group, which don't tell PP whether the Wall is on its left or its right. That is, of course, essential information for PP to "turn away from the Wall".

Associations of the other Groups use a variety of clues for the position of the Wall:

Group BA	Whether PP is touching the Wall on its left or right, and the position of the visible Stump.
Groups BB & BD	The direction robot A is facing and the position of the visible Stump.
Group BC	The position of the visible Stump.
Groups BG & BH	Whether PP is touching the Wall on its left or right.

The associations of Groups with high priority, like BA and BB, are more reliable but apply to fewer situations. The associations of low priority Groups, like BG and BH, apply to more situations but are more likely to be affected by other tasks.

Chapter 5

No Approval from Teacher

The interaction between robot A and robot B in chapter 4 was mostly driven by approval and disapproval from teacher through robot A. The aim was to teach PP (robot B) as fast as possible. In this shorter interaction of only 160 steps, there is no approval from teacher. Robot A never says ***Good*** and never **PATs,** but robot A did **SLAP** robot B once on step 60, and it said ***Bad*** 14 times. PP's motivation is from novelty goals and also from the 6 times it was approved automatically by Squashing the Cake.

Hand-Pointing is valid for this interaction. PP did point at the Cake on steps 71 and 72 when Pushing, but otherwise only at itself. That was all it saw robot A doing. Pointing to itself is represented in Appendix-4 by 'm' for 'me', while pointing at the Cake is 'o' for 'object'. Hand-Pointing goes with Speech, so it can't be shown in the graphical representation. I continue to omit Hand-Pointing in the discussion here, but the excerpt from the second interaction in Appendix-4 does show Hand-Pointing.

By imitating what it has seen and heard robot A do and say, PP learns to say what action it has just done. If PP says something inappropriate and robot A sees PP's action, robot A says ***Bad.*** Robot A sees robot B's action on only 75 of the 160 steps and says ***Bad*** 14 times. Robot B also has to see robot A's action for the imitation to work. Robot B does see robot A's action on 130 steps.

The following summary shows that there is steady learning during the interaction, with PP saying the correct statement on 28 steps in the first half, and on 64 steps in the second half:

An AGI Brain for a Robot. https://doi.org/10.1016/B978-0-323-85254-8.00005-3
© 2021 Elsevier Inc. All rights reserved.

PP moved FORWARD and said *I Moved Forward Q*: 62 times.

PP Pushed Cake and said *I Pushed Cake Q*: twice.

PP Squashed Cake and said *I Squashed Cake Q* after being told to: 3 times.

PP turned LEFT and said *I Turned Left Q*: 5 times.

PP turned RIGHT and said *I Turned Right Q*: 19 times.

PP touching Black Stump said *Look I Am Touching Black Stump Q*: once.

PP said the wrong statement after an action: 22 times.

PP said nothing or an incomplete statement after an action: 43 times.

PP failed to say the correct statement 45 times in the first half of the interaction, and 20 times in the second half.

Teacher, through robot A, spent the first 109 steps helping robot B to Squash the Cake by Pushing the Cake into a corner and by positioning robot A next to the Cake on the other side to where robot B appeared to be moving. From step 110 on, robot A leads robot B clockwise around the World to encourage it to make a long plan, which it can then follow. **Go on to the next section if you would like to skip the details.** This is how it went.

The World was started with the robots and Cake in the same positions as for the first interaction (Figure 7), so Reflex Motion takes robot B to Squash the Cake on step 3 in the NE corner of the World. A new Cake appears in the middle of the bottom row of the World. Robot A moves South to Push the Cake into the SW corner, and robot B follows it. Robot A moves around to position itself just above the Cake, so that PP (robot B) can Squash the Cake on step 14.

```
Step 3:                                                  Step 14:
        ---EC   3  ----E   3  ----E          -----   14  -----   14  -----
        -#--w B:F  -#--w A:L  -#--s          -#---  PP:F -#---  A:R  -#---
        ---#-      ---#-      ---#-          w--#-       w--#-       n--#-
        -----      --C--      --C--          CW---       W---C       W---C
```

The new Cake appears in the SE corner and robot A moves clockwise around the World until it is in the middle of the top row of squares on step 20. Robot B has followed robot A, so robot A now cuts down through the middle of the World to be on the far side of the Cake when robot B continues around the World. But robot B follows robot A down the middle of the World, so robot A Pushes the Cake up to the NE corner and sidles around to be on the other side of the Cake on step 34. Robot B reaches the Cake on step 36 but, instead of Squashing the Cake, robot B turns left on step 37. Robot A Squashes the Cake on step 37 and the Cake now appears at the left end of the second row.

```
--e--  20  N-e--  20  N-s--    ---nC  34  ---eC    ---eC  37  ---eC  37  ----e
N#---  PP:F -#---  A:R -#---    -#---  A:R -#---    -#--N PP:L -#--W  A:F C#--W
---#-       ---#-       ---#-    ---#-       ---#-    ---#-       ---#-       ---#-
----C       ----C       ----C    ----N       ----N    -----       -----       -----
```

Robot A heads off to the left to reach the Cake and Push it down into the SW corner. Robot B moves down to the SE corner and then turns back North and starts to move anticlockwise around the World. When on step 49, robot B turns to go down the middle of the World, robot A moves to block its progress, and robot B goes around the Black Stump to the SE corner from which it heads across to the Cake in the SW corner. In the meanwhile, robot A positions itself North of the Cake, ready for the Squash by robot B on step 62.

```
---W-  49  ---S-  49  ---S-        -----  62  -----  62  -----
-#---  B:L -#---  A:F -#---        -#---  PP:F C#---  A:L C#---
-e-#-       -e-#-       --e#-        s--#-       s--#-       e--#-
C----       C----       C----        CW---       W----       W----
```

The next Squash by robot B comes after only 11 steps. Robot A gets out of the way from being between robot B and the Cake, which has appeared at the left end of the second row. Robot A then WAITs below the NE corner while robot B Pushes the Cake up to the NW corner and along the top row to Squash the Cake on step 73.

```
---EC  73  ----E  73  ----E  74  ----S  74  ----S  75  -----  75  -----
-#--n PP:F -#-Cn  A:L -#-Cw PP:R -#-Cw  A:F -#Cw- PP:F -#CwS  A:F -#w-S
---#-       ---#-       ---#-       ---#-       ---#-       ---#-       --C#-
-----       -----       -----       -----       -----       -----       -----
```

After the Squash, the Cake appears in the square to the left of robot A. Robot A Squeezes the Cake against the Brown Stump on step 75, sending it South, and Squeezes it again against the bottom Wall on step 78, sending it East. By then robot B has moved down into the SE corner. Robot B doesn't attempt to Squash the Cake between the Stump and the Wall but moves back up North. Returning South, robot B Squashes the Cake on step 87.

```
-----  78  -----            -----  87  -----
-#---  A:F -#---            -#---  PP:F -#---
--s#-       ---#-            ---#S        ---#-
--C-E       --sCE            ---eC        --CeS
```

The Cake reappears on the bottom row next to robot A, so robot A Pushes the Cake along to the SW corner and then up to the NW corner. Robot B moves up to the NE corner and turns West on step 94, as if it intended to go across to the Cake, but then turns South, moves around the SE corner and across to the SW corner. Robot A, finding itself on the wrong side of the Cake for robot B's new approach, has to work its way around to the East side of the Cake in the NW corner. Robot B Squashes the Cake for the last time on step 109.

```
----N  94  ----W  94  ----W      Cw---  109  Nw---
C#---  PP:L C#---  A:W C#---     N#---  PP:F -#---
n--#-       n--#-      n--#-     ---#-       ---#-
-----       -----      -----     -----       ----C
```

Robot B moves South from the NW corner after Squashing the Cake, which reappears in the SE corner. After reaching the SW corner and turning around, robot B returns up to the NW corner on step 119. Robot A moves East to the NE corner and turns South, but after one step South it sees robot B returning to the NW corner and curves around to end up on the top row on step 119.

```
-----  119  N----  119  N--n-
N#-n-  PP:F -#-n-  A:F  -#---
---#-       ---#-       ---#-
----C       ----C       ----C
```

Both robots turn right to face East. Both robots move forward. When robot A sees that robot B is following it, robot A leads robot B around the World in a clockwise direction. On step 125, robot A reaches the Cake and starts to Push it around. By step 142, the robots have been right around the World with the Cake in front.

```
----E 125  ----S 125  ----S     ---E- 142  ----E 142  ----E 143  ----S 143  ----S
-#--- PP:R -#---  A:F -#---     -#--s PP:F -#--s  A:F -#--- PP:R -#---  A:F -#---
---#s      ---#s      ---#-     ---#C      ---#C      ---#s      ---#s      ---#-
----C      ----C      ---Cs     -----      -----      ----C      ----C      ---Cs
```

PP made a short plan at the end of step 146 and followed it. The long plan that PP makes on step 151 is followed almost to the end.

These plans will be discussed in the next section.

Planning in the Second Interaction

We hypothesized that a major function of consciousness was to plan for the future, allowing the organism to rapidly deal with many contingencies.

(Koch, 2004)

PP attempts to make a plan after every step. The first interaction was flooded with approval, so the plans found a goal and terminated very quickly. In the second interaction, there was no approval from robot A, so longer plans could be made. Long plans can be made if PP doesn't meet a goal. After step 125 of the second interaction, teacher made robot A lead robot B around the World to familiarize it with a route. This had the effect of removing novelty goals so that plans weren't terminated quickly by encountering those goals. The short plan that PP made on step 146 is shown in Figure 16.

This short plan is just 11 substeps long. (The "2-11" in Figure 16 means that PP is in the 2nd step of the plan and the 11th substep.) PP makes a much longer plan of 91 substeps on step 151, but the main ideas can be explained perfectly well with the short Plan. I continue to omit Hand-Pointing here for simplicity. However, the excerpt from the second interaction in Appendix-4 shows both plans with Hand-Pointing.

```
Step 146
PP: FORWARD              I Moved Forward Q
 A: FORWARD              I Pushed Cake Q
PLAN: 1-1    PP:RIGHT    I Turned Right Q
      1-6    A:FORWARD   I Pushed Cake Q
      2-11   PP:FORWARD              Plan ended with novelty goal. 11 substeps.
Step 147
PP:RIGHT                 I Turned Right Q
 A:FORWARD               I Pushed Cake Q
Step 148
 PP:FORWARD              I Moved Forward Q
 A:FORWARD               I Pushed Cake Q

----- 146  ----- 146  ----- 147  ----- 147  ----- 148  ----- 148
-#--- PP:F -#---  A:F -#--- PP:R -#---  A:F -#--- PP:F -#---  A:F
---#S      ---#-      ---#-      ---#-      ---#-      ---#-
--Cw-      --CwS      -Cw-S      -Cw-W      Cw--W      Cw-W-
```

Figure 16 A Short Plan on Step 146.

To begin the plan, the PP program makes a copy of Short Term Memory, which is called **STM-plan** (Recall chapter 2). STM-plan does not receive any events from the robot body and the World, but actions are chosen in the normal way by finding which action-predicting associations have all their context events in STM-plan. The highest priority associations of the BodyMove-predicting associations are 125BA and 126BB. These were first stored on step 129 and used several times since, so they have no novelty. Both predict a RIGHT turn so PP:RIGHT becomes the first BodyMove of the plan, as can be seen in Figure 16.

The RIGHT turn in the plan doesn't change the position of the robot in the World, as it would do if this was an ordinary RIGHT turn, but it does get put into STM-plan. Now, PP has to use its stimulus-predicting associations to find out what stimuli it should expect to receive after the RIGHT turn. The contexts used, for all 19 stimulus-predicting Groups, include the BodyMove that robot A has just done and robot A's position and direction, the positions of the Stumps, the position of the Cake and that it is the Cake, and what robot B is touching. There are also several predicted stimuli that tell it what stimuli robot A is receiving. Not all of these predictions are correct but they are enough for it to predict that PP will say

I Turned Right Q

in the plan. The predicted stimuli and the Words are fed into STM-plan.

The associations in Long Term Memory, which have been used to predict the actions and stimuli, are given 'plan marks'.

With STM-plan holding the most recent actions and stimuli of robot B, PP can now predict that robot A's next BodyMove would be FORWARD, as can be seen in the plan in Figure 16. Then PP predicts that robot A will say

I Pushed Cake Q

and STM-plan is again updated after each Word. The associations used have been given plan marks.

That completes the first step of the plan made at the end of the real step number 146.

When PP predicts its own next BodyMove for the plan, it finds that the association predicting FORWARD has a novelty mark in Long Term Memory. It has reached a goal and ends the plan! Now it is ready to **follow its plan**.

Following the plan begins on step 147 with Short Term Memory still holding the events that were in it when the plan was started. You can see from Figure 16 that the BodyMove and Speech actions on step 147 and the first half of step 148 are the same as those in the plan. PP is following its plan using Short Term Memory instead of STM-plan. PP checks that it is following the plan by noting and removing the plan marks that were placed in Long Term Memory while the plan was being constructed.

Wandering Through Long Term Memory

The basic conceptual distinctions assemble themselves into a scaffolding of meaning, which has hooks here and there on which to hang images, sounds, emotions, mental movies, and the other contents of consciousness.

(Pinker, 2007)

On step 151, a few steps after making the short plan, PP made a longer plan, which was 10 steps and 91 substeps long. This, again, was a very straightforward plan which was followed easily until the 9th step of the plan, when PP predicted wrongly that robot A would say

I Moved Forward Q

instead of

I Pushed Cake Q.

As a result, the plan-following ends on the 9th step with robot A's speech not matching its Words in the plan. (The curious can see the longer plan in the last excerpt of Appendix-4.)

Neither the short nor the longer plan are particularly interesting. They do little more than demonstrate the process of wandering through Long Term Memory by predicting the robot's own actions and then predicting the stimuli that would be received from the robot body and the World. With so little experience in its Long Term Memory, we cannot expect PP to make interesting plans.

These are simple plans made with simple rules. If plans were being made in the There-and-Then while the Here-and-Now behaviour was progressing, it wouldn't be so easy for PP to follow a plan, but plans could be used to strengthen paths through Long Term Memory in other ways. Associations could be strengthened and new associations could be formed.

My plan-making procedure is no more than a glimpse of what might be possible. We need only consider our own thinking and imagining to see how the future beckons.

Wandering through Long Term Memory could be used in many ways, such as planning a holiday, finding answers to questions, seeking solutions to problems, imagining doing something one hasn't done before, and just day-dreaming. In many cases, the wandering will start with only a partially relevant context in Short Term Memory, so predictions from more general, low priority associations may be needed to build up an adequate context for stronger predictions. This would be the case if, for example, PP was asked, out of the blue, for the meaning of a word.

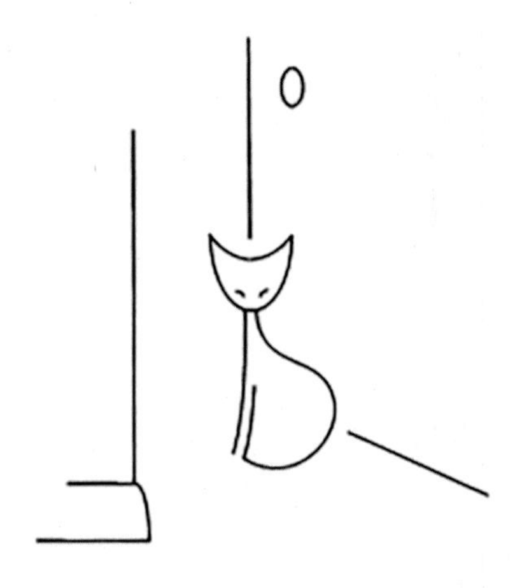

Chapter 6

Experiments from the Past

This chapter summarizes the ideas behind some of the experiments that have been carried out with PP in the past. All the computers available to me had memories of less than 1 megabyte. Most of the work was done on an EAI 640 digital computer that had 300 kilobytes for everything: the PP program and Long Term Memory! With that restriction, the number of Groups that could be used was minimal and they had to be chosen carefully. Also, I was interested in the basic design of PP, so the research was done with different kinds of Groups (called templates at that time). Choosing Groups becomes less of a problem when one can be generous with them.

Numbers in the Head

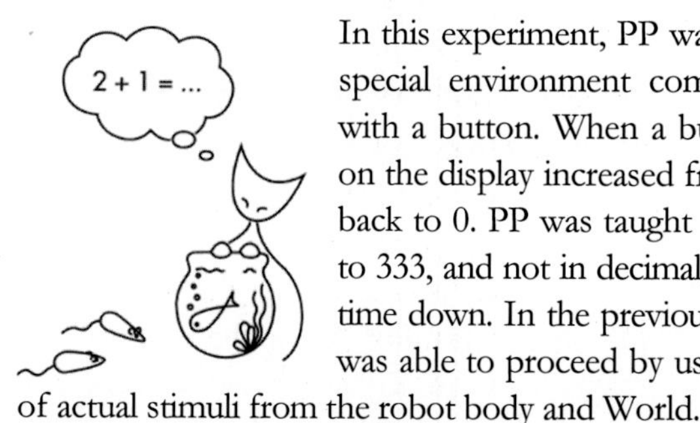

In this experiment, PP was first taught to count with a special environment comprising three displays, each with a button. When a button was pressed, a number on the display increased from 0 to 1 to 2 to 3 and then back to 0. PP was taught to count modulo 4 from 000 to 333, and not in decimal, in order to keep the learning time down. In the previous chapter, when planning, PP was able to proceed by using predicted stimuli in place of actual stimuli from the robot body and World.

In the second part of the counting experiment, predicted stimuli from the imagined counting are used to count dollar notes in a box. Many tasks require the assistance of learned techniques, like counting, carried out in the head. The details are given in chapter 4 of my 1998 book, *Associative Learning for a Robot Intelligence.*

An AGI Brain for a Robot. https://doi.org/10.1016/B978-0-323-85254-8.00006-5
© 2021 Elsevier Inc. All rights reserved.

Universal Turing Machine

Alan Turing's Universal Machine[51] is able to carry out any algorithm (computer program), so it is an important test of a system's computing ability. A major difficulty with teaching PP to emulate the Universal Turing Machine was boredom, as it would be for us. PP has a parameter, called Boredom, which increases if PP gets into a repetitive cycle (limit cycle) or trapping state. This has the effect of making PP's decisions random so as to give it the greatest chance of getting out of the limit cycle or trapping state. Teaching PP to emulate a Universal Turing Machine was a formal affair intended to show that PP has the computational power, for example, to learn language. The emulation is highly repetitive, so Boredom and novelty-seeking were switched off.

Marvin Minsky[52] invented a very neat Universal Turing Machine with a 7-state controller and 4-symbol vocabulary, which was ideal for a short demonstration. PP was **taught in parts**, which is a useful method enabling the whole system to be learned before emulation begins. The teacher has only to give a final command and away PP goes, emulating the Universal Turing Machine on its own and without stopping until it comes to a Halt command. The parts that had to be taught were:

1. the counting to enable PP to hold the whole of the tape of the Universal Turing Machine in PP's brain. Each number is a cell position on the tape and is the context of an association of a Working Memory Group. The most recent associated event of that association is the current symbol in that cell;
2. moving along the tape and reading and writing the symbols on it. Adding one to the count moves one position along the tape to the right. Subtracting one moves one position along the tape to the left;
3. the initial contents of the tape;
4. the quintuples (rules) of Minsky's machine; and
5. the ability to extend the tape whenever the end was reached.

This was an example of parallel processing because 5 parallel streams of actions were used. The first person to show that PP could emulate Minsky's Universal Turing Machine was Bruce MacDonald. His paper with me gave the details.[53]

The reader may well ask: But what about the PP being described here? Does it have Turing Machine power, which is needed to learn language?

Yes. This can be shown quite simply.

As it has to be done inside PP's brain, we can dispense with the World, do without stimuli, and do without Intonation. BodyMoves will be called Moves and will be used as internal controls: LEFT, RIGHT, HALT, QUINTUPLE, EXTEND, and XRIGHT. LEFT moves one cell to the left along the Tape by decreasing by one the number which says which cell is being looked at. RIGHT moves one cell to the right along the Tape by increasing the number by one. HALT stops the system. QUINTUPLE causes the quintuples of the Universal Turing Machine to be applied with the current State and Symbol. When a Symbol X is found on the Tape, meaning that the end of the Tape has been reached, EXTEND starts a sequence to add another cell to the Tape and move one cell to the right with XRIGHT.

We introduce a second Group for Working Memory, as it makes things a little simpler than using one Working Memory Group for all the memorizing. Hand-Pointing is used to distinguish Words for Symbols, States, and Counting. With this renaming of actions, the system can be taught the quintuples and the counting so that it emulates the Universal Turing Machine. It is a straightforward exercise.

PP wouldn't find this an interesting task unless rewarded regularly, but it has the ability in principle which is all that matters. We would also find it very difficult to remember the contents of the Tape when it was long.

Subroutines and Recursion

But it is also during childhood that theory of mind, episodic memory, and understanding of the future emerge. Childhood may be the crucible of the recursive mind. (Corballis, 2011)

The Universal Turing Machine is not something one would use in a program because it would be a very inefficient way of doing things. Subroutines are used all the time in programming and are a key to handling recursion. By reducing the well-known 15-puzzle to a 5-puzzle and using a hierarchical solution with Richard Korf's[54] micro-operators, we were able to present PP with a method for handling four (or more) levels of subtask. The main requirement for handling a subtask (or subroutine) is to remember, as you exit the subtask, what you were doing when you entered it, regardless of how many lower subtasks you have been in. An example would be a sentence with embedded clauses like

The man, *walked down the road.*
who held a book, *in his hand,*
which he was reading,

After completing the clause
which he was reading,
you have to remember that you were saying
who held a book
so as to continue *in his hand,*
and then you have to remember that you had started with
The man
so that you can end the sentence with *walked down the road.*

The reader may notice the similarity between the subroutine task and the Learning to Take Turns task of chapter 4. In each case, Working Memory is needed to hold information until it has been used and replaced by new information. In each task, there has to be a process for replacing the information in Working Memory and another process for reading the information that is in Working Memory.

In the Taking Turns case, the appropriate context (***Next Turn Is***) and Stress on the Word to be put in (<Mine or <Yours) were enough to store the Word. The Word could be read without Stress by using the context again.

The method used for entering and exiting the subtasks of the 5-puzzle did something rather similar to that for the Taking Turns. While the Taking Turns method used a Stressed Word for storing and unstressed for reading, the 5-puzzle method used a 'predicted sound' for storing and an ordinary sound for reading. To keep the storing and reading separate from the rest of the task, an **auxiliary action, RaiseEyeBrows,** was used in the associations concerned with storing in and reading from Working Memory. An auxiliary action is any action not used for carrying out a task. The details can be seen in our 1993 paper or in chapter 8 of my 1998 book.[55]

Auxiliary actions, like Intonation, Stress, and Raise Eyebrows, enabled PP to add structure to its learning.

Working Memory is so important that one can expect the human brain to have its own ways of implementing it. In the PP brain, any stored node can be used as Working Memory just by restricting reading to the associated event at the head of the list of associated events of the node: that associated event is the last associated event to have been stored with the context of the node. There must be many ways to achieve this, but it is not so easy to find a way that is straightforward to learn and to use. The Stress method used in Taking Turns is my best idea so far. We can expect the human brain to have many mechanisms for learning associations, updating them, and accessing stored associations.

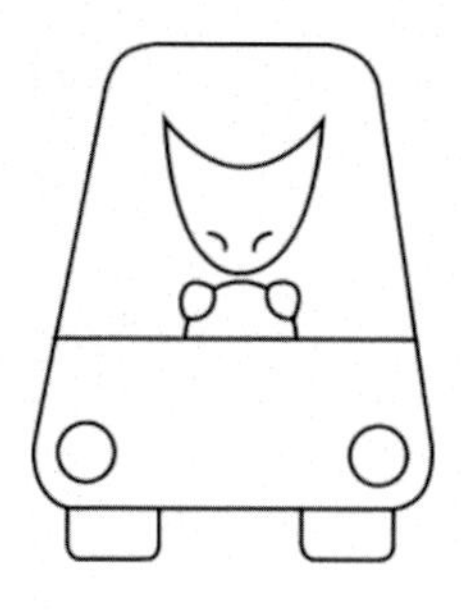

Boredom and Frustration

An experiment described in chapter 6 of my 1998 book attempted to repeat with PP an experiment reported by Elizabeth Bates[56] on the behaviour of 9-month old infants, regarding their ability to communicate intentions. She called it *the first moment*. My experiment used PP's plans to identify (or not) intentions. There is just one aspect of the experiment I need to mention here and that is to do with the making of plans.

During an interaction I had with PP many years ago, I made a mistake in my teaching. PP saw this as a novelty goal and kept making plans to get me to repeat the error, much to my annoyance. It showed me that there was a need for something to stop PP from getting into a plan-making cycle. The Boredom parameter, mentioned earlier, stops PP from going into a limit cycle in its ordinary behaviour, but not in the making of plans. I called the new parameter Frustration and it limited the repetitive making of plans to the same goal. In the copy of Elizabeth Bates' experiment, PP was made to 'cry' when Frustration hit its limit and stopped the planning.

I have avoided giving PP any features just for implementing emotions. They would add unnecessary complexity to the 'bare bones' system. Boredom and Frustration (without the crying) are designed to prevent system malfunction. When the PP brain is given a real robot body, emotions and expressions of emotion will become important for social interaction.[57]

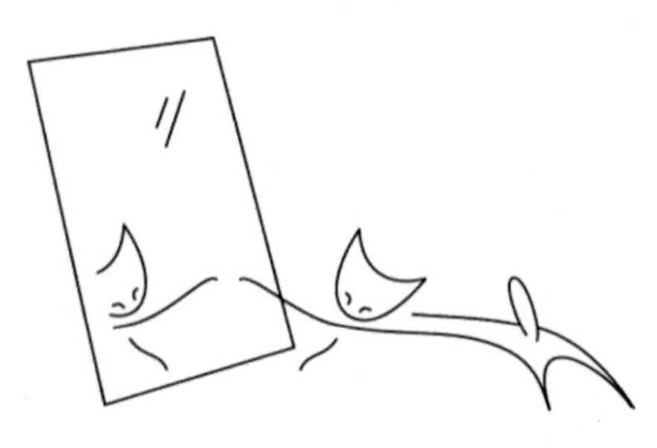

Chapter 7

Consciousness

When we marvel in those moments of heightened self-consciousness, at the glorious richness of our conscious experience, the richness we marvel at is actually the richness of the world outside, in all its ravishing detail. It does not "enter" our conscious minds, but is simply available. (Dennett, 1991)

How does the brain process environmental stimulation? How does it integrate information? How do we produce reports on internal states? These are important questions, but to answer them is not to solve the ***hard problem****: Why is all this processing accompanied by an experienced inner life?*

(Chalmers, 1996)

The information integration theory of consciousness (IITC) claims that a physical system has subjective experience to the extent that it is capable of integrating information. (Tononi, 2007)

In the end I still think that the hard problem is a meaningful conceptual *problem, but agree with Dennett that it is not a meaningful* scientific *problem.*

(Pinker, 2018)

All my thinking, however abstract, takes place in my sensory-motor space. Similarly, all PP's thinking takes place in the sensory-motor space, or event space, provided by PP's modules.[58]

When my computing resources no longer limited the amount of memory available for the programming of PP, I could have arranged for PP to make plans while it was moving about and speaking in its little World, had my programming skills been up to it. This would have raised an additional problem:

Could I allow PP to speak in a plan while speaking in its World?

There are a number of similar questions one can ask oneself. While doing this, can I at the same time think about that. While thinking about this, can I at the same time think about that?

While I am walking here on a smooth path, I have no difficulty in imagining walking in another part of the world.

An AGI Brain for a Robot. https://doi.org/10.1016/B978-0-323-85254-8.00007-7
© 2021 Elsevier Inc. All rights reserved.

While I am totting up some figures on a sheet of paper here, I am unable to imagine at the same time totting up a different set of figures in another house.

Talking and reasoning seem to require my conscious attention. Both are mental activities which use many different event-types in PP's Short Term Memory. If planning or wandering through Long Term Memory in the There-and-Then might happen at the same time as activity in the Here-and-Now, then the Here-and-Now must take priority. Planning and wandering would be able to occur only when PP is not talking or reasoning or otherwise using much of Short Term Memory in the Here-and-Now.

Perhaps PP could have a more flexible Short Term Memory with some parts behaving like STM-plan at times, so that it doesn't have to have separate Short Term Memories? It would then make sense to see Short Term Memory as a Global Workspace for the centre of consciousness. I haven't thought of a way to arrange that.

A straightforward way of deciding whether PP is conscious of something is to see whether PP can answer questions about it. If PP has associations that enable it to answer a variety of questions about that something, the answer is probably Yes. The questioning would not be in an examination or Turing Test, but would be over an extended period by people who had lived with the robot and had observed its development and experience. A great deal of learning will have to come before that is possible.

"What is the purpose of consciousness?" is another question that is often asked. The PP answer follows from the above test for consciousness: it is just the ability to think about and talk about our own and other people's decisions and plans.

It is difficult to talk about consciousness without imagining PP with a proper vision system. The vision PP had in the interactions of chapters 3–5 was minimal. We need to imagine something more adequate. When PP is put into a real robot its vision system will have to cope with all the complications of the real world.

Vision

This casts doubt on the idea that most neuroscientists and psychologists have, that we have an internal representation of the outside world. So that's one consequence of this way of thinking: that we don't re-present the world inside our brain. On the contrary, we use the outside world as a kind of outside memory to probe. There's no need to make an internal replica of the outside world. **(O'Regan, 2006)**

In chapter 10 of my 1998 book, I gave PP a special memory, called Painted Vision, to hold temporarily what PP's fovea was seeing as it moved around PP's World. If PP in a robot could be given a visual system comparable with ours, it would have to accommodate the scanning of the visual scene by the fovea, and Groups with delayed event-types could do the job, but the steps of the visual scanning would have to be much shorter than the substeps of the robot.

Suppose visual modules are presenting the PP brain with events of a number of visual event-types, such as lines, shapes, textures, colours, positions, movements, and shading. Groups of these event-types with delays of up to 7 or 8 foveal steps could then give associations with contexts that combine the snapshots of a foveal scan. Sequences of associated events from these associations could identify an object in a scene, see a route through a scene, or note a pattern of shapes and colours. As an example, imagine that the fovea is scanning around the painting robot cat:

Event-1: foveal snapshot of beret shape.
Event-2: foveal snapshot of face shape.
Event-3: foveal snapshot of scarf shape.
Event-4: foveal snapshot of arm and paintbrush shape.
Event-5: foveal snapshot of easel shape.

These events could form the contexts of associations with an associated event signifying "painting robot" for PP. In a similar way, one can imagine that a looped path in Long Term Memory could hold a meaning or concept by linking together different aspects and allowing entry at different points in the loop.

Our experience of seeing is magical. Don't we need a mechanism for bringing all the visual associated events together to form the integrated picture that we see with our eyes? Probably not! That question is the trap which treats the brain as a little homunculus which needs to have a picture to look at.

When all the high level, visual associated events are fed to the PP brain together with the eye and head movements, the brain will have the information it needs to integrate them into a whole scene. It doesn't need to "see" the whole scene because, like the light in the refrigerator which appears to be always ON because it is always ON when you look, every bit of the scene is immediately available when needed.

The PP brain should then, after much learning, be able to answer questions about its visual experience.

Belief Memory

Perhaps consciousness arises when the brain's simulation of the world becomes so complete that it must include a model of itself. (Dawkins, 1976)

Self-awareness is creating a model of the world and simulating the future in which you appear. (Kaku, 2014)

When PP makes a plan and then succeeds in following it, there is a sense in which PP has confirmed those associations that contributed to the plan. In my 2011 paper, I followed up that idea by giving PP a Belief Memory into which the confirmed associations could be stored and generalized. In the same paper, I gave PP a Trail Memory, which will be explained in a moment. The aim of the paper was to test David Rosenthal's HOT theory that a thought becomes conscious if it is the subject of a Higher Order Thought.[59] My rather unsatisfying result was that for the theory to be confirmed, "a thought" would have to be defined before the theory could be applied to intelligent machines.

Another theory is that of Philip Johnson-Laird,[60] who hypothesised that *Working memory is available only for those processes that yield conscious representations* and *Unconscious processes cannot use working memory.* The Working Memory that we have seen PP use (in earlier chapters) was part of Long Term Memory, so we can expect PP to be able to answer questions about its content but only after a great deal of learning.

Trail Memory

By contrast, 'the subjective present' is arguably the carrier and container of our conscious life, and everything that ever happens to us happens in it.

(Humphrey, 1992)

I had already had some sense of this when testing his memory, finding his confinement, in effect, to a single moment – 'the present' – uninformed by any sense of a past (or a future). Given this radical lack of connection and continuity in his inner life, I got the feeling, indeed, that he might not have *an inner life to speak of, that he lacked the constant dialogue of past and present, of experience and meaning, which constitutes consciousness and inner life for the rest of us. He seemed to have no sense of 'next' and to lack that eager and anxious tension of anticipation, of intention, that normally drives us through life.* (Sacks, 1995)

Looked at from the inside, consciousness seems continually to change, yet at each moment it is all of [one] piece – what I have called 'the remembered present' – reflecting the fact that all my past experience is engaged in forming my integrated awareness of this single moment. (Edelman, 2004)

A common human experience is like the following:

You open the French doors and walk out of the house into the garden. A green lawn stretches before you and you can see and hear a man pushing a motor mower from right to left in the far right corner. You turn right to head for the garden shed. Man and mower switch from the right side of your visual space to the left side. The mower continues moving, but in a different direction. You expect it to continue in this new direction. You are not fazed by this. As soon as you turn, your memory of the immediate past transforms into your current frame of reference. The mower appears to have been following a smooth trajectory, not leaping from one frame to another.

Trail memory does this for PP, including the present, the immediate past, and the expected immediate future, each remembered or anticipated in the frame of present experience. In my 2011 paper, *A Multiple Context Brain for Experiments with Robot Consciousness,* Trail Memory was provided as an "extended present" for Bernard Baars' Global Workspace theory of consciousness. Nicholas Humphrey[61] calls it 'thick time'. Belief Memory and Trail Memory enabled me to have PP modify its own plans, which it did in a small way. This was to be a Higher Order Thought, but, as said above, it raised the question: What is a thought? Belief Memory and Trail Memory can be seen as future directions for the development of PP.

Feelings and Perceptions

The experienced quality of a sensory feel derives from a sensorimotor interaction that involves the particular "what it's like" aspects of richness, bodiliness, insubordinateness, and grabbiness. The extent to which an agent is conscious of such an experience is determined by the degree of access the agent has to this quality. ... Finally, if the agent has some sort of notion of self, then it begins to make sense to say that the agent is really consciously experiencing.

(O'Regan, 2011)

Feelings and perceptions (called 'qualia' by philosophers) are sensations that you experience when you see the colour red, hear a violin, touch a stinging nettle, smell a rose, or taste a pineapple. These sensations enable you to distinguish red from blue, violin from cello, stinging nettle from oak leaf, rose from lavender, and pineapple from apricot. We cannot tell whether our own sensations that give us these discriminations are the same as other people's sensations for the same discriminations, but we can guess that they are similar. If we found that someone else could make exactly the same discriminations down to the finest detail, then it would be more likely that they were having the same sensations, but no test could tell us exactly what they felt, because feelings and perceptions are private.

Does it matter?

No, it is only the ability to make the discriminations that matters.

Yes, because we can make physical devices that can distinguish colours or sounds or smell or whatever, that we know couldn't have feelings or perceptions because they work by simple physical processes and have no brains.

To have feelings and perceptions, you have to be conscious of those sensations. So the big question is not "What are feelings and perceptions?" but "What does a brain have to have in order to be aware of the discriminations it uses?" Answer the second question and you will know that the brain has sensations, but you will still not know what other people's and robots' sensations are like!

When I get a jab of pain in my big toe, I wiggle my back because I know that the nerve from my big toe to my brain is being squeezed in my spine. In one of the houses we lived in, there were ceiling beams and I had a loop of rope that enabled me to pull myself up and stretch my spine so as to release the squeezed nerve. All my brain needs in order to tell me that my toe is hurting is a neural signal which it assumes to be from my big toe. All my seeing and sensing modules inform my brain through signals on the axons of neurons. If these neural signals don't solve Chalmers' 'hard problem' and give us an 'inner experienced life' (see quotation at the beginning of this chapter), why should we expect additional neural signals to do so?

Aside-5 Of course, PP would have to be given pain stimuli to feel (be aware of) pain.

If you provide the PP brain in a robot body with a scent event-type, which has different events for every scent that you want PP to be able to distinguish, and a nose module, which can input to the brain a different event for each of those scents, then PP will be able to distinguish all those scents. Also, it will be able to learn to answer questions about scents, in the same way that it was learning to distinguish touching the Brown Stump and touching the Black Stump in the section Talking About Touching in chapter 4.

Will PP have a 'smell feeling' for scents?

What kinds of questions could we ask PP in order to find out if it was aware of smelling scents?[62]

My guess: If (1) the PP brain is given a body which provides it with the same sensory input as my body provides me, down to the finest detail (same hues of colours, same visual pattern details, same timbers of tones, same sensations of touch, same odours, pains, etc.), (2) PP can make the same distinctions between sensations as I can, and (3) PP can discuss with me the differences that we can discriminate, then the sensations will feel the same to PP as they do to me.[63]

I think this is more or less what Daniel Dennett[64] was saying when he dismissed qualia in his book *Consciousness Explained,* so I am not claiming any originality for my guess. It seems also to be compatible with David Chalmers'[65] principle of organizational invariance.

In his deep and delightful book *Soul Dust: The Magic of Consciousness,* Nicholas Humphrey (2011) has difficulty in explaining how sensation acquires its *phenomenal richness* and its *thickness of time.* He considers reentrant feedback, attractor states, and strange loops. Here is a different possibility.

When the PP brain is put in a real robot body with good vision, it will encounter Dennett's *richness of the world outside, in all its ravishing detail.* It will see a sunset and there will be an **explosion of novelty goals.** PP will go on looking (transfixed?) as it finds goal after goal. After becoming fluent with language much later, PP, in Humphrey's own words for a humanoid robot, *would make all the right claims about the soul-hammering qualities of its experience, ineffability, privileged access, and so on.*[66]

Conscious Robots

The sort of consciousness that is essential for creativity, because it is involved in the very definition of the term, is self-reflective evaluation. A creative system must be able to ask, and answer, questions about its own ideas. (Boden, 1990)

On the delusional theory of consciousness the answer is simple. If any machine had language or memes or whatever it takes to be able to ask the question "Am I conscious now?" and concoct theories about its inner self and its own mind, then it would be as deluded as we are and think it was conscious in the same deluded way. (Blackmore, 2005)

It might be added that, in my opinion, the openness of the physical world is needed to explain – rather than explain away – human freedom.

Karl Popper in (Popper & Eccles, 1977)

In my 1987 Metascience article,[67] *Design of a Conscious Robot,* I developed a series of definitions of the behaviour of a robot with the aim of being able to determine whether a robot was conscious or not. A robot satisfying a definition was called a **yammy** ("I am me") robot. The fifth and final definition can be seen as a distant goal of the project to develop PP:

> Definition: A robot will be described as yammy-5 when its behaviour in an <u>open</u> environment is most efficiently predicted in terms of
>
> (i) its own intentions,
>
> (ii) its knowledge of the intentions of other purposeful entities,
>
> (iii) its knowledge of their knowledge of its intentions,
>
> (iv) its ability to predict its own and other entities' future intentions,
>
> and
>
> (v) the answers it gives to its own questions about its own intentions.

For "open environment" read "real world". This may seem to be rather a lot to expect of a robot, but I was thinking of the engineering problem of having a robot doing a job in a remote and hazardous environment where humans cannot go and humans cannot communicate with the robot in real time.

For example, the task might be on the ocean floor or in space, possibly with other robots. Before the robot goes, one would want to know what the robot intended to do. When it came back, an account would be expected of how the task went, how it coped with unforeseen events, and how it would deal with similar events on another occasion.

When the PP brain is put into a real robot body, it will have a lot to learn about the real world. There will have to be considerable research on modules needed to turn information provided by the sensors in its robot body into the most useful event-types for the PP brain.[68] It will need a generous provision of Groups. Perhaps Deep Learning can be used in modules for segmenting speech into words and for providing good vision for PP.

Language and learning in a world with more than one other agent should enable a well-endowed PP to become aware of itself and other agents.

PP will need to learn about the physical world. If this box is pushed it will move but if that wall is pushed it stays put. This is a ball and it will roll on a flat surface. That is a lever and can be used to move heavy objects. PP will have to learn about agents that do things, say things, and express their feelings by actions and facial expressions. PP will have to learn the meanings of words and sentences. PP will have to learn the difference between self and other, between *mine* and *yours*, between kindness and cruelty, and between truth and falsity.

PP will become a yammy-5 robot.

What a difference good vision and manipulable hands will make!

Best wishes to anyone who puts the PP brain in a real robot!

There is a long way to go.

Main Points

Multiple Context Associative Learning is computational.

PP is like a river with no controlling computer program.

PP is controlled by its learned associations.

Novelty goals give PP free will.

PP can learn from single exposures to events, such as Words.

Working Memory can be implemented in different ways.

PP can learn and use language.

Auxiliary actions enable PP to add structure to its learning.

PP can plan and wander in thought through its Long Term Memory.

We will not know if PP is aware of its sensations until we can discuss the question with PP.

PP needs a real robot body.

Notes

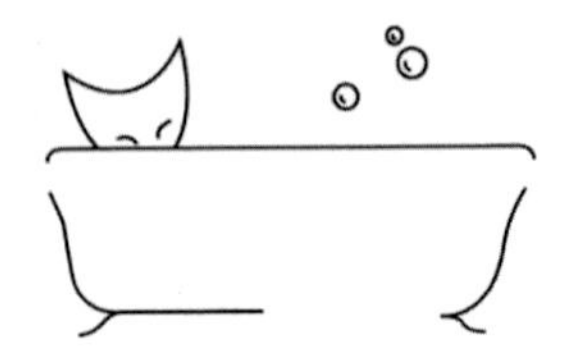

---

[1] Baum (2004, page 168). Another example is from Pinker (2002, pages 40, 41): *Humans behave flexibly because they are programmed: their minds are packed with combinatorial software that can generate an unlimited set of thoughts and behavior.*

[2] Mithen (1996, page 17) describes the last 6 million years as a play in which: *Our ancestors are the actors, their tools are the props and the incessant changes of environment through which they lived the changes of scenery.*

[3] The last significant change in the body was probably the gradual alteration to the larynx and vocal tract that made speech possible and distinguished the human brain from those of the other apes less than a million years ago. The change to bipedal movement happened well before that with Lucy (Australopithecus afarensis) *adept at both walking upright on two legs and climbing trees* (Mithen, 1996, page 19).

[4] The modules are similar to those imagined by Fodor in his book *The Modularity of Mind* (1983) although Fodor didn't have modules taking high level commands from the central brain to the muscles of the body (motor programs). The central brain receives high level information from the output of each Fodor module, no information coming directly to it from stages of processing within the module, so the modules are said to be 'encapsulated' or 'cognitively impenetrable'. Some maturation and learning can occur within a module, resulting in new events appearing at the output over time.

[5] PP is an example of Reinforcement Learning which is a big subject with powerful mathematical techniques, as can be seen in the classic text and its new 2nd edition (Sutton & Barto, 1998, 2018). The LeakBack process is a suboptimal, conditional probability method of helping PP to reach its nearest goals. In chapter 4, PP's goals include the built-in goal of Squashing the Cake, the volatile goals set up by a PAT action and the "Good" Word from teacher, negative goals from SLAP and "Bad", and the just as important volatile novelty goals. However, none of these goals are directly related to success of the PP system, which may loosely be described as an increase in knowledge and intelligence. There are biological reasons for favouring the LeakBack method, as explained in Note 39, but there must be room for improvement overall.

[6] When a novelist uses a word processor to write a novel, all the story and the ideas for the novel come to the word processor by input from the novelist. The word processor may correct spelling and even improve the grammar. We could describe the word processor as an 'input-driven program', but it wouldn't be helpful or accurate to call the word processor the author of the novel. Also, we wouldn't want to say that the novelist was just the input to a computer program.

[7] The acronym PURR-PUSS, from which PP is derived, stood for **P**urposeful **U**nprimed **R**ewardable **R**obot with **P**redictors **U**sing **S**lides and **S**trings, because of the original method of implementing the PUSS associations. PUSS is now better described by **P**redictors **U**sing **S**hort **S**egments of context. It was shortened to **PP** in my 1998 book. Andy Clark (2016, page 1) has recently called his theoretical 'predictive processing' models PP, which fits well with my PP as it could also be called a predictive processor.

[8] It would be a mistake to call the learning program (PP program), with its input of associations, the brain. The mistake would be closely analogous to calling a word processor the author of a novel. In each case the mistake attributes responsibility to the wrong system. The changing collection of associations (PP brain) is responsible for the brain's 'thinking', not the relatively fixed learning program (PP program) that has evolved over thousands of years to provide the brain with the best information from the body by administering modules, event-types and Groups. The learning program is an 'experience processor'. It holds the experience obtained from interaction between the body and the real world. A word processor holds the novel that is being written by the novelist. In chapters 4 and 5, the behaviour of the robot is described in terms of the associations that PP has stored by interacting with its little World. The associations are the brain of the robot. Another analogy is between the learning program and a computer's operating system. The changing collection of associations can be seen as a collection of rules, so the collection can be seen as a program of sorts implemented by the learning program acting as an operating system. In this case, therefore, the brain is a program, separate from the learning program, but only the learning program has evolved over thousands of years. The brain 'program' is changing continuously as it is acquired by interaction between the body and the real world.

[9] The Prologue of my 1977 book included these words: *The Waimakariri River winds down from the Southern Alps across the Canterbury Plain to the Pacific Ocean. On a hill above the river stands a small man-made house with a water system comprising an arrangement of pipes and tanks. The water system of the house, like so many human designs, sets out to prescribe precisely each twist and turn that the water in the house must take. The mistakes of the designer are*

revealed by air-locks and blockages, by overflows and leakages. The water in the river obeys the universal rule of gravity and flows downhill. It is not stopped by dams and other obstacles – only delayed. The general trend in artificial intelligence research today is to design systems that are more and more complex. In my analogy, the typical system is a house with a multitude of pipes and tanks. ... We can expect quick spectacular results from the "pipe" programs of artificial intelligence, but, in my view, their achievements will be superseded and surpassed by "river" programs, like PURR-PUSS. The recent successes of AI with Deep Learning can be seen as supporting my prediction and restoring respect for associative learning.

[10] Gödel (1931, 1962) (Turing, 1936). The PP brain is a collection of associations, which are being modified or added to on every step, so it can't be a fixed program as required by Gödel and Turing. Of course, if learning was halted the associations would then be equivalent to a program, but learning is never halted.The Gödel argument was elaborated extensively by Roger Penrose in his books *The Emperor's New Mind* (1989) and *Shadows of the Mind* (1994). He sought ways to avoid these limitations, but overlooked the solution I had used for PP and ended up trying to explain the human mind with quantum mechanics. Michael Arbib wrote in 1987, page 183: *Gödel's theorem limits a human as much as a machine; ... So, any really adequate cybernetic theory of thought must contain a full understanding of growing inductive machines that continually change their internal structure as a result of interacting with the external environment.*

[11] Andreae (1977, page 6).

[12] Excellent accounts of the brain's complexity are by Churchland & Sejnowski (1992, 2017) and by Arbib, Erdi, & Szentagothai (1998). A recent account for the general reader is the New Scientist Instant Expert: *How Your Brain Works.* (Williams, 2017)

[13] Evolution has been able to produce reliable modules for us because it has had hundreds of thousands, if not millions, of years to do so. Designing modules for a robot body will be no easy task, but it will be helped by the increasing knowledge from neuroscience of what information the human visual cortex, in particular, feeds into the central brain.

[14] A comprehensive account of Deep Learning is by Goodfellow, Benjio, & Courville (2016). See also *The Deep Learning Revolution.* (Sejnowski, 2018)

[15] Deep Learning is a statisticial technique that processes large amounts of data, whereas the PP brain's associative learning has a computational base (associations, which are rules, and Groups) with a statistical overlay (transitions

and Leakback). The PP brain can learn from single data items, which is difficult for Deep Learning. Deep Learning could be used to implement modules for the PP brain. Dennett (2017, page 399) writes *My view is (still) that deep learning will not give us --- in the next fifty years --- anything like the "superhuman intelligence" that has attracted so much alarmed attention recently.*

[16] PP was designed to have the computational strength needed for language at a time when the memory size of computers was too small to allow the simulation of elaborate robot vision and real robots weren't available. The inputs and outputs of associations were assumed to be ON/OFF rather than graded. A change from binary (ON/OFF) to vector coding (Churchland & Sejnowski, 1992-2017) would make an enormous difference to PP's vision but the logical strength of binary coding may still be needed for language.

[17] Andreae & MacDonald (1991) used nursery rhymes to confirm that George Miller's (1956) magic number 7 $\pm$ 2 was about right for the maximum length of a sound sequence in a context.

[18] For example, Bernard Baars (1997) who uses a theatre metaphor with *the spotlight of attention shining on the stage of working memory.* More recently, Stanislas Dehaene (2014, page 163) wrote: *We have sketched a specific theory of consciousness, the 'global neuronal workspace', that is the condensed synthesis of 60 years of psychological modeling.*

[19] My first learning machine in 1963, called STeLLA, did not have the computational power to learn language. See also Gaines & Andreae (1966). PP was designed to have that power (Andreae & Cleary, 1976). The work for my first two books was done on computers with less than a megabyte of memory. New results presented here have been obtained on my modern personal computer with gigabytes of memory.

[20] Associative learning has had its detractors, the most notable being Noam Chomsky, who demolished B. F. Skinner's attempt to explain language by behaviourist associative learning (Chomsky, 1959). Since then the standard version of language learning has been with Chomsky's proposal that we are born with a Universal Grammar (UG). Recently, philosophers Jerry Fodor and Zenon Pylyshyn (2015, page 12) have said *"This transition from associative to computational accounts of cognitive processes has the look of a true scientific revolution.* Multiple Context Associative Learning is both associative and computational, so the Fodor-Pylyshyn revolution makes no sense. Chomsky (2016, page 21) has written: *To deny the existence of UG---that is, of a biological endowment underlying the capacity for language---would be to hold that it is a miracle that humans have language but other organisms do not*. Berwick & Chomsky (2016) give a comprehensive argument for UG, but accept that it could

be compatible with *some slight rewiring of the brain.* (page 110). Perhaps this is no more than an increase in delayed event-types. There has been growing disbelief in an innate Universal Grammar, unique to humans. Many people, including Elman, et al. (1996), Tomasello (2003), Arbib (2012), Evans (2014) and Corballis (2017), favour a gradual evolution of language from the signs, primitive sounds and facial expressions of primate ancestors of humans. In the short interactions carried out so far with PP, there hasn't been a need to distinguish syntax from semantics. The Subroutines and Recursion section in chapter 6 refers to the use of a single 'auxiliary action' to give structure to a problem, so perhaps actions from semantics can build up syntactic structures and integrate syntax with semantics.

[21] *At first, infants' memory for operant responses is highly context dependent. If 2–6-month olds are not tested in the same situation in which they were trained—with the same mobile and crib bumper and in the same room—they remember poorly. This specificity of infant memory also applies to imitation; young babies will imitate adult-modeled actions on a toy only when given a toy identical in color and features to the one the adult used. After 9 months, the importance of context declines* (Berk, 2013, page 142). The perceived 'decline' is probably explained by the learning of supporting contexts (nodes) as discussed at the end of the "How Much has PP Learned?" section in chapter 4 of my book.

[22] Hofstadter (2007).

[23] Dennett (1984).

[24] Steven Pinker (2002), in his book *The Blank Slate,* attacked empiricism, romanticism and dualism, which he identified with the doctrines of the Blank Slate, the Noble Savage and the Ghost in the Machine. PP is quite different from the kind of system envisaged by the behaviourists and avoids romanticism and dualism, but sits comfortably in the empiricism chair!

[25] The main task that I am teaching PP is the holding of information in Working Memory. It is something that Steven Pinker said associative learning couldn't do. He wrote (2002, page 21): *For example, storing the value of a variable in the brain, as in "x = 3", is a critical computational step in navigating and foraging, which are highly developed talents of animals in the wild. But this kind of learning cannot be reduced to the formation of associations, and so it has been ignored in neuroscience.* The remembering numbers task in chapter 4 was inspired by this remark.

[26] According to the New Scientist Instant Expert: *How Your Brain Works* (Williams, 2017, page 12), information enters and leaves the cortex through about a million neurons.

[27] Previously, Groups were called Templates but that title seems to have given people the idea that Templates were restrictions imposed on the brain, rather than being additional facilities discovered by evolution.

[28] To find, update or store an association of a Group in Long Term Memory, the PP program runs down a list of events for the first event-type of the Group to find an event that matches the corresponding event in Short Term Memory. If it finds a matching event, that event puts it on the list for the second event-type, and so on for each event-type of the Group. If the event from Short Term Memory is not on a list, the PP program adds it to the list and starts a new list for the next event-type. When it reaches the last event-type of the Group, the matching event will, or will be made to, point to an event in the list of associated events of the node, which can be read or updated as required. The lists of associated events are in Long Term Memory. The computer science technique of 'hashing' would be a faster method of storing associations, but this 'tree' method is easier for debugging the PP program and for 'forgetting'. A Short Term Memory conveyor belt (shift register; or first in, first out buffer) that I have used in the past moves one step forward only if it gets a non-null event. Its events are called 'threading events' and it can reach further back in time.

[29] Arbib, Erdi, & Szentagothai (1998) describe a different organization: *STM puts together instances of schemas drawn from an LTM that encodes a lifetime of experience in a vast network of interconnected schemas* (p. 44). (STM and LTM are Short and Long Term Memory.)

[30] Assuming only 50 event-types to represent sight, sound, touch, taste, smell and proprioception, and a maximum of only 12 event-types in any Group, the total number of possible Groups is $50C_{12} + 50C_{11} + \ldots + 50C_5 + 50C_4 + 50C_3 + 50C_2 = 172,186,125,405$.

[31] A popular account of the plasticity of the brain is given by Doidge (2007).

[32] With intense practice, as when a pianist develops control of both hands doing different movements at the same time, or a guitar player adds voice to his or her performance, it may be that new Groups are formed, but I think that it is more likely that the integration, speed and smoothness of these skills is done by the cerebellum. Ryan & Andreae (1993) showed how PP could be coupled to Albus' CMAC cerebellum model to enable PP to control smoothly a rolling ball in a maze by tilting the maze. Norman Doidge (2007, page 67) offers another explanation. *When a child learns to play piano scales for the first time, he tends to use his whole upper body – wrist, arm, shoulder – to play each note. ... With practice ... he develops a "lighter touch," ... the child goes from using a massive*

number of neurons to an appropriate few, well matched to the task. Perhaps strong associations take over from a mass of weaker associations.

[33] Learning at two levels must take place at two time scales. Imagine for example that you were learning the words of a language, but at the same time a new spelling dictionary was being issued annually. New spellings like 'cyse' for 'cheese', 'catte' for 'cat', 'weall' for 'wall' and 'straet' for 'street', kept appearing. (Winchester, 2003, page 6). Imagine the confusion! Fortunately, the spellings of words usually change over centuries while the learning of the words themselves happens in a lifetime.

[34] McCulloch & Pitts (1943) showed that neurons could perform logical operations, which are basic to computation. A neuron version of PP would be a better model of how the brain learns and might reveal more clever and flexible equivalents of PP's Groups. An electronic neural version would not have neurotransmitters sensitive to externally applied and internally generated drugs.

[35] Bickhard & Terveen (1996) give a detailed account of encodingism and say on page 283 *The central point to be made, however, is that connectionism and PDP approaches are just as committed to, and limited by, encodingism as are compositional, or symbol manipulational, approaches (though in somewhat differing ways).*

[36] Strictly, any cluster of Groups with associated event-types which never occur together on the same step can be assigned a network, but this is rare. The PP program allows for such clusters.

[37] An important use of the most recent associated event can be seen when PP is emulating a Universal Turing Machine and its tape in chapter 6. For each numbered cell on the tape, the most recent associated event will say what symbol is in the cell at that time.

[38] The LeakBack calculations use Ronald Howard's (1960) discounted value iteration method. They use the conditional probabilities held by the transitions between nodes. Details are given in my 1998 book.

[39] An early suggestion of how LeakBack might be implemented in a biological brain by leaking back a chemical, such as RNA or protein, was made in the appendix to my 1977 book. The New Scientist Instant Expert "How Your Brain Works" says: *Many neuromodulators act on just a few neurons, but some can penetrate through large swathes of brain tissue creating sweeping changes. Nitric oxide, for example, is so small (it's the 10th smallest molecule in the known universe, in fact) that it can easily spread away from the neuron at its source. It alters receptive neurons by changing the amount of neurotransmitter released*

with each nerve impulse, kicking off the changes that are necessary for memory formation in the hippocampus (Williams, 2017, page G).

[40] Goals and LeakBack provide PP with an 'attention mechanism'. Novel events of all kinds produce novel associations which are marked as novelty goals, until they are found again. Other goals are formed by approval of various kinds. LeakBack guides PP's activity towards the nearest and strongest of the goals, and away from negative goals. But this is unlikely to be the whole story. Susan Blackmore (2005, page 40) wrote: *In 1890, William James famously proclaimed that 'Every one knows what attention is', but many subsequent arguments and thousands of experiments later it seems that no one knows what attention is, and there may not even be a single process to study.*

[41] This is a simple account with many details omitted. Some of these will get mentioned when particular applications are considered. Others can be found in my 1998 book. Having now described the structure and operation of the PP system, I need to reiterate that PP has a simple brain, in no ways competing with the magnificent efforts of those unravelling the detailed operation of the human brain. I also need to distance PP from statements like this one on page 92 of Arbib (2012): *The rest of the cortex is called association cortex, but this is a misnomer that reflects an erroneous 19th-century view that the job of these areas was simply to 'associate' the different sensory inputs to provide the proper instructions to be relayed by the motor cortex. The absolutely false idea that 90% of the brain is 'unused' probably arose from a layman's misinterpretation of the fact that the exact functions of many of these 'association areas' were very little known until the last few decades.* PP is not using simple 19th-century associations. PP's Groups recognize the functions of the many association areas. Every baby has an as yet unfilled, enormous, memory capacity.

[42] A small blemish on step 995 was that PP said *Q* instead of *CQ*. The Q and CQ endings have been omitted here.

[43] A program error, which prevented robot B from seeing what robot A was pointing at, was due to a wrong definition of the Hand-Pointing event-type. It was corrected for the second interaction described in chapter 5.

[44] For mirror neurons see Ramachandran (2000, 2011), Arbib (2006) and Arbib (2012).

[45] In earlier work, I used a reflex mechanism called "echo-speech" in my 1977 book and "mimic-sound" in my 1998 book. If PP heard a sound and hadn't made a sound itself, the heard sound was stored as if PP had made it itself. Then it tended to make sounds as it had heard others saying them. We do seem to have a tendency to complete speech sequences to ourselves or even aloud, when

someone starts a familiar sequence. Imitation stores the other robot's speech with the context in which it was spoken, while echo-speech and mimic-sound store the other robot's speech in the hearing robot's context. Imitation is more logical but I have wondered whether there is still a place for my old method. Because I don't take PP through the babbling stage, the only way that PP can acquire a new word is through imitation (or mimic-sound). Once it has an association with the new word as associated event, PP can use the new word in other contexts through generalization and exploration.

[46] My colleague, Koenraad Kuiper (1980), posed a challenge for PP in the form of a sentence: *It was her Mini which the police believed Andrea to have been driving, wasn't it?* Kon explains the difficulties in understanding the sentence and in particular the number of words that have to be kept in Working Memory to understand later parts of the sentence. The experiments being described here show that PP can carry information along in Working Memory.

[47] It is worth noting that in the interactions the robot body and the World are simulated but the PP brain is real. My brain in the other robot is real too! Seven excerpts from the first interaction are given in Appendix-4 as well as an excerpt from the second interaction. Both interactions and the PP program are available from ResearchGate on the internet. The excerpts are steps 1-15 (start), 225-240 (taking turns), 577-595 (Cake to Roll), 645-676, 693-716 (remembering numbers), 945-972 (touching) and 989-997 (ending). One excerpt from the second interaction is steps 129-160 (planning).

[48] For the record, PP's responses to the 134 commands 'You Turn Right' were 100 correct with 2 unblocked, 17 wrong and 17 no responses. Responses to the 115 commands 'You Turn Left' were 87 correct with 4 unblocked, 11 wrong and 17 no responses. Responses to the 178 commands 'You Move Forward' were 167 correct with 4 unblocked, 6 wrong and 5 no responses. Responses to the 89 commands 'You Push Cake' were 87 correct and 2 no responses. Responses to the 18 commands 'You Squash Cake' were 17 correct with 1 no response. After 421 FORWARD actions by PP, it said *I Moved Forward* 203 times, *I Squashed Cake* 16 times, *I Pushed Cake* 33 times, *I Pushed Roll* 5 times (2 with prompting), nothing 49 times, other messages 67 times and an incorrect statement or nonsense 48 times. After 177 RIGHT turns by PP, it said *I Turned Right* 111 times, nothing 27 times, other messages 22 times and an incorrect statement or nonsense 17 times. After 175 LEFT turns by PP, it said *I Turned Left* 122 times, nothing 11 times, other messages 22 times and an incorrect statement or nonsense 20 times.

[49] If you look carefully at Figure 10, you will see that Group SE is missing HearSelfOrOther(-1), Group SF is missing HearSelfOrOther(-2) and Group SG is

missing HearSelfOrOther(-3). The associations of these Groups allow a new Word to be accepted in place of a known Word if the rest of the context is the same. In this way, PP can accept the change from Cake to Roll without being disrupted. It is an example of the "slot and filler" mechanism for learning simple analogies, e.g. page 165 of Michael Tomasello's *Constructing a Language (*2003).

[50] The associations can be understood using Figure 10 and the following abbreviations:

N = North relative to the robot i.e. straight ahead.
42 = Square adjacent to robot on right.
U = unknown. Wl = Wall. Rb = Robot. Sp = Space. PtMe = Point at Self.
HearFO = HearSelfOrOther. TchF = TouchFront. VOthPos = OtherRobotPosn.
ObPos = ObjectPosn. VBn/k = BrownStumpPosn/BlackStumpPosn.

[51] Turing (1936).

[52] Minsky (1967).

[53] MacDonald & Andreae (1981) and chapter 5 of Andreae (1998). This demonstration was important because Pat Hayes had said in 1977 that PP could only learn finite automata *so it certainly couldn't learn a grammar for any reasonably interesting subset of English.*

[54] Korf (1985).

[55] Andreae, Ryan, Tomlinson, & Andreae, P. M. (1993) and chapter 8 of Andreae (1998).

[56] Bates (1979), chapter 2, page 34.

[57] Igor Aleksander's (2005, page 31) five axioms *on which the design and functioning of a materially conscious machine can be based* include being part of the world, being aware of past experience, being purposeful and thinking ahead, all of which fit comfortably with PP. However, his fifth axiom, of doing what is determined by *feelings, emotions and moods,* differs from PP's doing what is determined by its own and given goals. Mark Lee (2020) writes on page 243: *As we are interested in creating useful robots, and not trying to model complete simulations of human states and behavior, we do not need to build emotional or human values into our robots.* Certainly, robots will not need emotions that have evolved solely for human procreation.

[58] The event space generated by the modules is expanded into the much much larger context space by PP's Groups. The context space should be able to support Gilles Fauconnier's (1985) 'mental spaces' and the more complex processes of 'conceptual blending' (Fauconnier & Turner, 2002).

[59] My 2011 paper is *A Multiple Context Brain for Experiments with Robot Consciousness*. Rosenthal's HOT theory is in his 2005 book *Consciousness and Mind.*

[60] Philip Johnson-Laird hypothesized about Working Memory in his 2006 book *How We Reason.*

[61] Humphrey (2006).

[62] At the end of Dan Dennett's (1994) discussion of the M.I.T. humanoid robot Cog, he wrote: *In fact, I would gladly defend the conditional prediction: <u>if</u> Cog develops to the point where it can conduct what appear to be robust and well-controlled conversations in something like a natural language, it will certainly be in a position to rival its own monitors (and the theorists who interpret them) as a source of knowledge about what it is doing and feeling, and why.*

[63] If you think that PP could be a zombie which answered all my questions without being aware of different sights, sounds, smells, etc., then you will not accept my guess! We aren't aware of all our vision, but we cannot describe the vision that we aren't aware of. This includes the saccades our eyes make and the remarkable examples of subconscious dorsal stream vision reported by Goodale & Milner (2004). If PP were given some subconscious vision, it would not go to Short Term Memory and would not become available to PP's associations. In the case of actions, PP supports Benjamin Libet's famous experiment (see Libet, 1999) showing that an action is chosen before becoming consciously available. In PP, an action is chosen before it goes into Short Term Memory and then, via Groups, into Long Term Memory to become associations available for conscious thought. As Julian Jaynes (1976 page 39) put it, *one does one's thinking before one knows what one is to think about.*

[64] Dennett (1991, ch.12).

[65] Chalmers (1996, pages 248-9).

[66] Humphrey (2011, pages 52, 72, 88). Dennett's words are from the quotation at the head of the chapter.

[67] Andreae (1987) and chapter 7 of Andreae (1998). All animals that sense and react to their environments may be said to have a level of subconscious awareness. Conscious awareness, or knowing that you are aware, may begin with animals that can pass the mirror test for self-awareness. Being able to discuss and answer questions about one's awareness is a stronger criterion for human awareness. The PP brain is <u>implemented</u> in a computer or other hardware, not simulated; only its body and its world, excluding me, are simulated. The Integrated Information Theory of Consciousness of Giulio Tononi

(Koch, 2019, page 81) requires that a conscious system has intrinsic causal power *That is, its current state must be influenced by its past and it must be able to influence its future.* PP has intrinsic causal power, most obviously shown by its ability to plan. Perhaps, in a sufficiently richly simulated body and world, PP could show conscious awareness, but it wouldn't be worth the trouble to find out: it would be easier to use a real robot.

[68] The design of modules should take into account the abilities that a baby is born with. Lake, Ullman, Tenenbaum, & Gershman (2016) write *A different picture has emerged that highlights the importance of early inductive biases, including core concepts such as* ***number, space, agency, and objects,*** *as well as powerful learning algorithms that rely on prior knowledge to extract knowledge from small amounts of training data.* We can expect the need for modules that present the brain with objects and their movements, facial characteristics like a pair of eyes to identify agents and the robot's own eye movements to map out space.

Index to Main Text

References

Aleksander, I. (2005). *The World in my Mind, My Mind in the World.* Imprint Academic. p.31.

Andreae, J. H. (1963). STeLLA: A Scheme for a Learning Machine. *Proc 2nd IFAC Congress.* (pp. 497-502). Basel: Broida, ed, Automation and Remote Control, Butterworths 1969.

Andreae, J. H. (1977). *Thinking with the Teachable Machine.* Academic Press.

Andreae, J. H. (1987). Design of a Conscious Robot. *Metascience vol.5*, 41-54.

Andreae, J. H. (1998). *Associative Learning for a Robot Intelligence.* Imperial College Press.

Andreae, J. H. (2011). A Multiple Context Brain for Experiments with Robot Consciousness. *IEEE Trans. Autonomous Mental Development vol.3 no.4*, 312-323.

Andreae, J. H., & Cleary, J. G. (1976). A New Mechanism for a Brain. *Int. J. Man-Machine Studies vol.8 no.1*, 89-119.

Andreae, J. H., & MacDonald, B. A. (1991). Expert Control for a Robot Body. *Kybernetes vol.20*, 28-54.

Andreae, J. H., Ryan, S. W., Tomlinson, M. L., & Andreae, P. M. (1993). Structure from Associative Learning. *Int. J. Man-Machie Studies, vol.39*, 1031 - 1050.

Arbib, M. A. (1987). *Brains, Machines and Mathematics.* Springer-Verlag 2nd edn.

Arbib, M. A. (2006). *Action to Language Via the Mirror Neuron System.* Cambridge Univ. Press.

Arbib, M. A. (2012). *How The Brain Got Language; The Mirror System Hypothesis.* Oxford Univ. Press.

Arbib, M. A., Erdi, P., & Szentagothai, J. (1998). *Neural Organization.* MIT Press.

Baars, B. (1997). *In the Theater of Consciousness.* Oxford Univ. Press. p.42.

Bates, E. (1979). *The Emergence of Symbols.* Oxford Univ. Press.

Baum, E. B. (2004). *What Is Thought?* MIT Press. p.168.

Berk, L. E. (2013). *Child Development.* Pearson. p.142.

Berwick, R. C., & Chomsky, N. (2016). *Why Only Us: Language And Evolution.* MIT Press.

Bickhard, M. H., & Terveen, L. (1996). *Foundational Issues in Artificial Intelligence and Cognitive Science.* Elsevier.

Blackmore, S. (2005). *Consciousness: A Very Short Introduction.* Oxford Univ. Press. pp.40,132.

Bloom, P. (2000). *How Children Learn the Meanings of Words.* MIT Press. p.10.

Boden, M. A. (1990). *The Creative Mind.* Cardinal. p.279.

Boden, M. A. (2016). *AI: Its Nature and Future.* Oxford Univ. Press. p.56.

Chalmers, D. (1996). *The Conscious Mind.* Oxford Univ. Press. p.xi.

Chomsky, N. (1959). Review of "Verbal Behavior" by B.F.Skinner. *Language vol.35 no.1*, 26-58.

Chomsky, N. (2016). *What Kind of Creatures Are We?* Columbia Univ. Press.

Churchland, P. S., & Sejnowski, T. J. (1992, 2017). *The Computational Brain.* MIT Press. pp.9,163.

Clark, A. (2016). *Surfing Uncertainty.* Oxford Univ. Press.

Clark, H. H. (1996). *Using Language.* Cambridge Univ. Press. p.387.

Corballis, M. C. (2011). *The Recursive Mind.* Auckland Univ. Press. p.202.

Corballis, M. C. (2017). *The Truth about Language.* Auckland Univ. Press.

Cover. (2019). The robots are ready. All they need now is a brain. *New Scientist*, February 2.

Dawkins, R. (1976). *The Selfish Gene.* Oxford Univ. Press. p.63.

Dehaene, S. (2014). *Consciousness and the Brain.* Viking Penguin.

Dennett, D. C. (1984). *Elbow Room.* Oxford Univ. Press. p.170.

Dennett, D. C. (1991). *Consciousness Explained.* Penguin Press. p.408.

Dennett, D. C. (1994). Consciousness in Human and Robot Minds. *IIAS Symposium on Cognition, Computation and Consciousness.* Kyoto.

Dennett, D. C. (2017). *From Bacteria to Bach and Back.* W. W. Norton & Co.

Dennett, D. C. (2018). Facing up to the hard question of consciousness. *Phil. Trans of Royal Society B 373:20170342. http//dx.doi.org/10.1098/rstb.2017.0342.*

Deutsch, D. (2019). Beyond Reward and Punishment. In J. Brockman, *Possible Minds* (pp. 113-124). Penguin Press.
Doidge, N. (2007). *The Brain that Changes Itself.* Viking Penguin.
Edelman, G. M. (2004). *Wider Than the Sky.* Yale Univ. Press. p.8.
Elman, H. L., Bates, E. A., Johnson, M. H., Karmiloff-Smith, A., Parisi, D., & Plunkett, K. (1996). *Rethinking Innateness.* MIT Press.
Evans, V. (2014). *The Language Myth.* Cambridge Univ. Press.
Fauconnier, G. (1985). *Mental Spaces.* MIT Press.
Fauconnier, G., & Turner, M. (2002). *The Way We Think.* Basic Books.
Fodor, J. A. (1983). *The Modularity of Mind.* MIT Press.
Fodor, J. A., & Pylyshyn, Z. W. (2015). *Minds without Meanings.* MIT Press.
Gaines, B. R., & Andreae, J. H. (1966). A Learning Machine in the Context of the General Control Problem. *Proc 3rd Int. Fed. Automatic Control IFAC Congress.* Paper 14B. London: Institute of Mechanical Engineers.
Gödel, K., translated by Meltzer, B., & Introduction by Braithwaite, R. B. (1931, 1962). *On Formally Undecidable Propositions of Principia Mathematica and related Systems.* Oliver & Boyd.
Goodale, M. A., & Milner, A. D. (2004). *Sight Unseen.* Oxford Univ. Press.
Goodfellow, I., Benjio, Y., & Courville, A. (2016). *Deep Learning.* MIT Press.
Gopnik, A. (2019). AIs Versus Four-Year-Olds. In J. Brockman, *Possible Minds* (pp. 219-230). Penguin Press.
Greenfield, S. (2006). Conversation. In S. Blackmore, *Conversations on Consciousness* (pp. 92-103). Oxford Univ. Press.
Hayes, P. (1977). Small Brains and Large Minds. *AISB European Newsletter, April, no.26*, 10-14.
Heaven, D. (2017). *New Scientist Instant Expert: Machines That Think.* John Murray Learning. p.222.
Hofstadter, D. (2007). *I Am A Strange Loop.* Basic Books. p.340.
Howard, R. A. (1960). *Dynamic Programming and Markov Processes.* MIT Press.
Hume, D. (1740, 1978). *A Treatise of Human Nature, ed L.A. Selby-Bigge.* Oxford Univ. Press. p.93.
Humphrey, N. (1992). *A History of the Mind.* Chatto & Windus. p.170.
Humphrey, N. (2006). *Seeing Red.* Harvard Univ. Press.
Humphrey, N. (2011). *Soul Dust: The Magic of Consciousness.* Quercus.
Jaynes, J. (1976). *The Origin of Consciousness in the Breakdown of the Bicameral Mind.* Houghton Mifflin Company.
Johnson-Laird, P. N. (2006). *How We Reason.* Oxford Univ. Press.
Kaku, M. (2014). *The Future of the Mind.* Doubleday. p.57.
Koch, C. (2004). *The Quest for Consciousness.* Roberts & Co. p.323.
Koch, C. (2019). *The Feeling of Life Itself.* MIT Press.
Korf, R. E. (1985). *Learning to Solve Problems by Searching for Macro-operators.* Pitman.
Kuiper, K. (1980). PURR-PUSSful Language Learning. *Man-Machine Studies: UC-DSE/16 ISSN 0110 1188*, pp. 5-10.
Lake, B. M., Ullman, T. D., Tenenbaum, J. B., & Gershman, S. J. (2016). Building machines that learn and think like people. *Behavioral and Brain Sciences no.24, e253.*
Lee, M. H. (2020). *How to Grow a Robot.* MIT Press.
Libet, B. (1999). Do We Have Free Will? In B. Libet, A. Freeman, & K. Sutherland, *The Volitional Brain.* (pp. 47-57). Imprint Academic.
Locke, J. (1690, 1947). *An Essay Concerning Human Understanding*, ed Raymond Wilburn. J. M. Dent & Sons. p.26.
MacDonald, B. A., & Andreae, J. H. (1981). The Competence of a Multiple Context Learning System. *Int. J. of General Systems, vol.7*, 123-137.
McCulloch, W. S., & Pitts, W. (1943). A Logical Calculus of the Ideas Immanent in Nervous Activity. *Bulletin of Mathematical Biophysics, vol.5*, 115-117.

Miller, G. A. (1956). The Magical Number Seven, Plus or Minus Two: Some Limits on Our Capacity for Processing Information. *Psychological Review, vol.63, no.2,* 81-96.
Miller, G. A., Galanter, E., & Pribram, K. H. (1960). *Plans and the Structure of Behavior.* Holt, Rinehart & Winston. p.111.
Minsky, M. L. (1967). *Computation: Finite and Infinite Machines.* Prentice-Hall.
Mithen, S. (1996). *The Prehistory of the Mind.* Thames & Hudson.
O'Regan, J. K. (2006). Conversation. In S. Blackmore, *Conversations on Consciousness* (pp. 160-172). Oxford Univ. Press.
O'Regan, J. K. (2011). *Why Red Doesn't Sound Like a Bell.* Oxford University Press.
Penrose, R. (1989). *The Emperor's New Mind.* Oxford Univ. Press.
Penrose, R. (1994). *Shadows of the Mind.* Vintage, Oxford Univ. Press.
Pinker, S. (2002). *The Blank Slate.* Penguin Books.
Pinker, S. (2007). *The Stuff of Thought.* Viking Penguin. p.82.
Pinker, S. (2018). *Enlightment Now.* Penguin Books. p.427.
Popper, K. R., & Eccles, J. C. (1977). *The Self and the Brain.* Springer-Verlag. p.51.
Ramachandran, V. (2000, 2011). Mirror Neurons and Imitation Learning as the Driving Force Behind "the Great Leap Forward" in Evolution. In J. Brockman, *The Mind* (pp. 101-111). Harper Perennial.
Ramakrishnan, V. (2019). Will Computers Become Our Overlords? In J. Brockman, *Possible Minds* (pp. 181-191). Penguin Press.
Rosenthal, D. M. (2005). *Consciousness and Mind.* Oxford Univ. Press.
Ryan, S. W., & Andreae, J. H. (1993). Learning Sequential and Continuous Control. *Proc. 1st NZ ANNES Conf.* ed Nikola K. Kasabov (pp. 302-305). IEEE Computer Society Press.
Sacks, O. (1995). *An Anthropologist on Mars.* Picador. p.46.
Sacks, O. (2017). *The River of Consciousness.* Picador. p.25.
Sejnowski, T. J. (2018). *The Deep Learning Revolution.* MIT Press. p.158.
Sutton, R. S., & Barto, A. G. (1998, 2018). *Reinforcement Learning,* 1st & 2nd edns. MIT Press.
Tomasello, M. (2003). *Constructing a Language: A Usage-Based Theory of Language Acquisition.* Harvard Univ. Press.
Tononi, G. (2007). The Information Integration Theory of Consciousness. In M. Velmans, & S. Schneider, *The Blackwell Companion to Consciousness* (pp. 287-299). Blackwell Publishing.
Turing, A. M. (1936). On Computable Numbers, with an Application to the Entscheidungsproblem. *Proc. London Math. Soc. series 2, vol. 42,* 230-265.
Velmans, M. (2006). Conversation. In S. Blackmore, *Conversations on Consciousness* (pp. 233-244). Oxford Univ. Press.
Williams, C. (2017). *New Scientist Instant Expert: How Your Brain Works.* John Murray Learning.
Winchester, S. (2003). *The Meaning of Everything.* Oxford: Oxford University Press.

Appendix-1

The PP program

1 **Robot B.** Obtain BodyMove for robot B:
Make Choice List of BodyMoves and choose Best.
If B can't choose, use random motion to choose BodyMove.
Shift new BodyMove into shift register in Short Term Memory (STM).
Do BodyMove In World and update Long Term Memory (LTM) with BodyMove.
Obtain all stimuli from World, except Sound and Pointing, shift into STM and update LTM.

2Obtain Speech (including Intonation and Stress) for robot B.
Make Choice List of Speech actions and choose Best.
If B can't choose, Speech becomes DRP-Quiet without Stress.
Shift new Speech actions into STM. Do Speech in the World. Update LTM with Speech actions.
Obtain all Speech stimuli from World, shift into STM and update LTM.

Obtain Hand Pointing for robot B:
Make Choice List of Hand Pointing actions and choose Best.
If B can't choose, Pointing becomes Null-Point.
Shift new Pointing action into STM. Do Pointing in the World. Update LTM with Pointing action.
Obtain all Pointing stimuli from World, shift into STM and update LTM.

If Intonation is FLT, YES or NEG, repeat from **2**.
If Intonation is RIS, swap to robot A and start with Speech.
If Intonation is DRP or CMD, do LeakBack, and try to make a Plan .
If Plan succeeds, start Plan-following with BodyMove of robot A.
If Plan is unsuccessful, start Bodymove with robot A and no plan-following.

3 **Robot A.** Obtain BodyMove action from robot A. Do BodyMove In World.
4 Obtain Speech(including Intonation and Stress) from robot A. Do Speech in the World.
Obtain Hand Pointing from robot A. Do Pointing in the World.
If Intonation is FLT, YES or NEG, repeat from **4**.
If Intonation is RIS, swap to robot B and start with Speech at **2**.
If Intonation is DRP or CMD, swap to robot B and start with BodyMove from **1**.

-----

update Long Term Memory (LTM) with new action
For each Group with action type as its associated event type: find nodes with contexts in STM.
For each node: put at, or bring to, the front a transition with the new action.
Update probability and destination of transition of last node of Group with current node.
Update approval and disapproval parameters of current node.

The Java program was written with BlueJ and it is available on ResearchGate.
When the program runs it creates a number of data files:
Main file is a series of files of 500 steps, each with details of the interaction.
Response Out file lists the teacher's inputs and enables the interaction to be rerun.
Summary file gives a summary of the interaction without details.
Sparse file gives the output which can be seen in Appendix-4.

Architecture

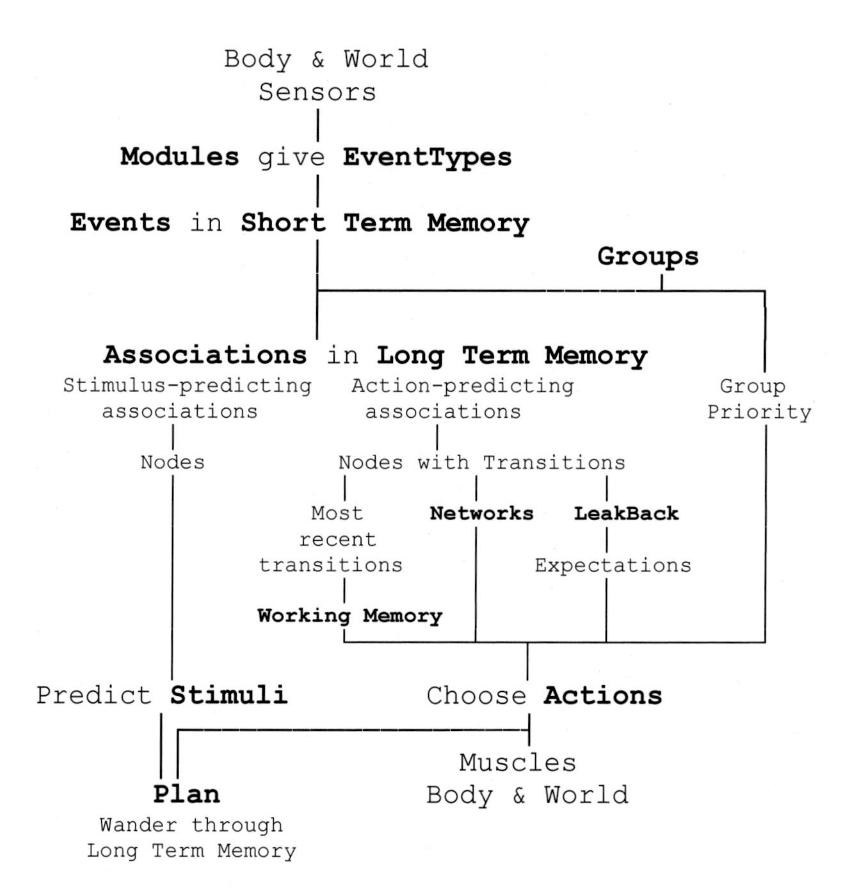

Appendix-2

Squashes in First Interaction

Step	Squashed by	Corner	Orientation	Wrote to WM for next turn	Read WM for this turn
3	B	NEcnr	cBA	No	No
17	A	NWcnr	cBA	No	No
36	PP	SWcnr	cBA	No	No
60	A	SWcnr	cAB	Yes	No
88	PP	NWcnr	cAB	No	Prompt
104	PP	SWcnr	cBA	Yes	Yes
116	A	SEcnr	cBA	Yes	Prompt
139	PP	SEcnr	cAB	Yes	Prompt
150	A	NWcnr	cBA	Yes	Yes
165	PP	NEcnr	cAB	Yes	Prompt
179	A	SWcnr	cAB	Yes	Prompt
200	PP	NEcnr	cAB	Yes	Yes
228	A	NEcnr	cBA	Yes	Unblocked
242	PP	NWcnr	cBA	Yes	Yes
255	A	NEcnr	cAB	Yes	Yes
267	PP	SWcnr	cAB	Yes	Yes
289	A	SWcnr	cBA	Yes	Yes
303	PP	SWcnr	cAB	Yes	Wrong
316	A	SEcnr	cBA	Yes	No
341	PP	SEcnr	cBA	Yes	No
350	A	NWcnr	cAB	Yes	No
370	PP	NEcnr	cAB	Yes	Unblocked
390	A	NEcnr	cBA	Yes	Yes
404	PP	NWcnr	cBA	Yes	Yes
421	A	SEcnr	cAB	Yes	Yes
435	PP	NEcnr	cBA	Yes	Prompt
446	A	SWcnr	cAB	Yes	Unblocked
465	PP	NEcnr	cAB	Yes	Yes
474	A	SWcnr	cBA	Yes	Yes
492	PP	SEcnr	cBA	Yes	Yes
505	A	SWcnr	cAB	Yes	Yes
518	PP	SEcnr	cBA	Yes	Unblocked
527	A	NWcnr	cAB	Yes	Yes
569	PP	NEcnr	cAB	Yes	Yes
580	**B**	NWcnr	cAB	Yes	Yes
594	A	SWcnr	cBA	Yes	Yes
636	A	SEcnr	cBA	Yes	Yes
683	PP	SWcnr	cAB	Yes	Yes
803	A	NEcnr	cAB	Yes	Yes
996	PP	NEcnr	cAB	Yes	Yes

WM = Working Memory

Appendix-3
Touching

The robots' phrases are:

Look I am touching Brown Stump.
Look I am touching Black Stump.
What am I touching?

Look you are touching Brown Stump.
Look you are touching Black Stump.
What are you touching?

The directions North, East, South, and West give the side of the Stump which is being touched.
The robot A or B is the robot saying the phrase.
The numbers give the steps on which the phrase was said when the robot was on the side of the Stump given by the direction.
Brackets around numbers indicate that the phrase said was inappropriate.
Robot A has said each phrase from almost every side of both Stumps.
Robot B hasn't learned to say "What am I touching?" at all.

Robot A From direction:

Brown Stump	**North**	**East**	**South**	**West**
Look I am touching:	858	748-9,754,762	774,793,843,846	831
What are you touching?	740,778	859	758,835-6	791
Look You are touching:	776,789	841,846	761,833	792
What am I touching?	860	-	781	-

Black Stump				
Look I am touching:	733,889,897	741-2	723,897,969	728,869,880
What are you touching?	812,930	879,955	893,901,906	-
Look You are touching:	806,813,866, 915,933,972	877	887	750,763
What am I touching?	891,910,913-5	816	936,939,966,972	870,872,948, 950,952

Robot B From direction:

Brown Stump	**North**	**East**	**South**	**West**
Look I am touching:	-	(822),859-61	758,(831),836()	792
What are you touching?	-	-	846	(866)
Look You are touching:	-	-	-	-
What am I touching?	-	-	-	-

Black Stump				
Look I am touching:	929-30,933,936, 939,965-6,969,	947-8,955,958	723,901,905-6, 909-10,913-4	- 972
What are you touching?	-	-	-	-
Look You are touching:	915	-	936,939,966	950,952
What am I touching?	-	-	-	-

Postscript: An extension of the first interaction to 1046 steps shown in ResearchGate on the internet has robot B using each of the phrases correctly at least once.

Appendix-4

Excerpts from the Interactions

The description of the robot's World and the meanings of actions, etc. are given in chapter 3. After each four steps, a graphic representation of the steps is shown. Comments are given on the right side of the page. All Hand Pointing is omitted from the first interaction.

A BodyMove of robot A is shown by "A:", of robot B using Reflex Motion (RM) by "B:", and of the PP brain through robot B by "PP:". "^" shows an action was prompted.

Bold is used for special actions by robot A.

First Interaction - Steps 1-15

1
B:FORWARD Q
A:RIGHT Next Turn Is Yours So You Move Forward CQ

2
B:FORWARD Q
A:WAIT Next Turn Is Yours So You Squash Cake CQ — A WAITs to show PP so can't Pat.

3
B:FORWARD (Squashed) Q — B is automatically approved.
A:LEFT You Squashed Cake So Next Turn Is <Mine Q — A shows B how to set up **WM**.

4
B:RIGHT Q
A:FORWARD I Moved Forward Q — B saw A move.

```
-E--C   1  --E-C   1  --E-C   2  ---EC   2  ---EC   3  ----E   3  ----E   4  ----S   4
-#--w  B:F -#--w  A:R -#--n  B:F -#--n  A:W -#--n  B:F -#--n  A:L -#--w  B:R -#--w  A:F
---#-      ---#-      ---#-      ---#-      ---#-      ---#-      ---#-      ---#-
-----      -----      -----      -----      -----      --C--      --C--      --C--
```

5
B:FORWARD Q
A:FORWARD I Moved Forward Q — B saw A move.

6
PP:FORWARD Q — PP makes its first move.
A:LEFT You Move Forward CQ

7
B:FORWARD Q
A:PAT You Turn Right CQ

8
B:RIGHT Q
A:PAT You Move Forward CQ

```
----S   5  -----   5  -----   6  -----   6  -----   7  -----   7  -----   8  -----   8
-#-w-  B:F -#-wS  A:F -#w-S PP:F -#w--  A:L -#s--  B:F -#s--  A:P -#s--  B:R -#s--  A:P
---#-      ---#-      ---#-      ---#S      ---#S      ---#-      ---#-      ---#-
--C--      --C--      --C--      --C--      --C--      --C-S      --C-S      --C-W
```

9
B:FORWARD Q
A:PAT You Push Cake CQ

10
B:FORWARD Q
A:RIGHT You Push Cake CQ — A was too busy to Pat.

11
PP:FORWARD Q
A:RIGHT I Turned Right Q — A was too busy to Pat.

12 — B didn't see A move.
PP:FORWARD Q
A:FORWARD I Moved Forward Q — B didn't see A move.

```
-----   9  -----   9  -----  10  -----  10  -----  11  -----  11  -----  12  -----  12
-#s--  B:F -#s--  A:P -#s--  B:F -#s--  A:R -#w-- PP:F -#w--  A:R -#n-- PP:F -#n--  A:F
---#-      ---#-      ---#-      ---#-      ---#-      ---#-      ---#-      C--#-
--C-W      --CW-      --CW-      -CW--      -CW--      CW---      CW---      W----
```

13
B:LEFT I Turned Right CQ
A:LEFT Bad You Turn Right CQ

14
B:RIGHT Q
A:FORWARD You Turn Right CQ

15
PP:RIGHT Q
A:PAT You Push Cake CQ

=================================

First Interaction - Steps 225-240

```
225
 PP:RIGHT Q                                           B was told to turn Right and PP did so.
 A:WAIT You Push Cake CQ                                             A is out-of-reach to Pat.
226
 PP:FORWARD Next Q
 A:WAIT You Push Cake CQ                                             A is out-of-reach to Pat.
227
 PP:FORWARD Next Turn Is Yours So You Move Forward CQ        PP reads WM,tells A to move with
 A:FORWARD Good Next Turn Is Mine So You Push Cake CQ                             unblocking.
228
 PP:FORWARD Next Turn Is Yours So You Squash Cake CQ           PP reads WM,tells A to Squash
 A:FORWARD (Squashed)  Good I Squashed Cake So Next Turn Is <Yours Q        with unblocking.
NC--- 225  EC--- 225  EC--- 226  -EC-- 226  -EC-- 227  --EC- 227  --EC- 228  ---EC 228
-#--- PP:R -#---  A:W -#--- PP:F -#---  A:W -#--- PP:F -#---  A:F -#--n PP:F -#--n  A:F
---#n      ---#n      ---#n      ---#n      ---#n      ---#n      ---#-      ---#-
-----      -----      -----      -----      -----      -----      -----      -----
229                                                              PP turned Left towards Wall.
 PP:LEFT You Squashed Cake So Next Turn Is <Mine Q                  PP sets up WM with <Mine.
 A:SLAP Good You Turn Left CQ
230
 PP:LEFT I Turned Left Q
 A:LEFT Good You Move Forward CQ                                        A is too busy to Pat.
231
 PP:FORWARD I Moved Forward Q
 A:PAT Good You Turn Left CQ
232
 PP:LEFT I Turned Left Q
 A:PAT Good You Push Cake CQ
---En 229  ---Nn 229  ---Nn 230  ---Wn 230  ---Ww 231  --W-w 231  --W-w 232  --S-w 232
-#C-- PP:L -#C--  A:S -#C-- PP:L -#C--  A:L -#C-- PP:F -#C--  A:P -#C-- PP:L -#C--  A:P
---#-      ---#-      ---#-      ---#-      ---#-      ---#-      ---#-      ---#-
-----      -----      -----      -----      -----      -----      -----      -----
233
 PP:FORWARD I Pushed Cake Q
 A:PAT Good You Push Cake CQ
234
 PP:FORWARD I Pushed Cake Q
 A:PAT Good You Squeeze Cake CQ
235
 PP:FORWARD Next Q
 A:FORWARD You Turn Right CQ                                         A is out-of-reach to Pat.
236
 PP:RIGHT Q
 A:PAT You Push Cake CQ
--S-w 233  ----w 233  ----w 234  ----w 234  ----w 235  ----w 235  ---w- 236  ---w- 236
-#C-- PP:F -#S--  A:P -#S-- PP:F -#---  A:P -#--- PP:F -#---  A:F -#--- PP:R -#---  A:P
---#-      --C#-      --C#-      --S#-      --S#-      ---#-      ---#-      ---#-
-----      -----      -----      --C--      --C--      -CS--      -CS--      -CW--
237-2392
 PP:FORWARD Bad You Squash Cake CQ                                            B can't see A.
 A:FORWARD Bad You Push Cake CQ                                         A is too busy to Pat.
238-2404
 PP:FORWARD Next Turn Is Mine^ So Q
 A:WAIT You Turn Right CQ                                            A is out-of-reach to Pat.
239-2416
 PP:RIGHT Q
 A:WAIT You Push Cake CQ                                               A must Wait so no Pat.
240-2423
 PP:FORWARD Next Turn Is Mine^ So You Squash Cake CQ
 A:WAIT Bad Next Turn Is Yours So You Push Cake CQ
---w- 237  ---w- 237  --w-- 238  --w-- 238  --w-- 239  --w-- 239  --w-- 240  --w-- 240
-#--- PP:F -#---  A:F -#--- PP:F -#---  A:W -#--- PP:R -#---  A:W -#--- PP:F C#---  A:W
---#-      ---#-      ---#-      C--#-      C--#-      C--#-      C--#-      N--#-
-CW--      CW---      CW---      W----      W----      N----      N----      -----
           =================================
```

Cake Replaced by Roll. Steps 577-595

577
PP:FORWARD I Moved Forward Q
A:FORWARD I Squeezed Roll Q
A didn't see B.
B saw A Squeeze.
578
PP:RIGHT Good You Turn Right CQ
A:RIGHT Next Turn Is Mine So You Turn Left CQ
579
PP:LEFT Q m
A:FORWARD Next Turn Is Mine So Now You Wait CQ
580
B:FORWARD (Squashed) I Squashed Roll^ So Next Turn Is <Yours Q
Unblocking and prompt.
A:RIGHT Good You Squashed Roll So Next Turn Is <Mine Q

```
--W-- 577  -W--- 577  -W--- 578  -N--- 578  -N--- 579  -W--- 579  RW--- 580  W---- 580
-#--- PP:F -#---  A:F R#--- PP:R R#---  A:R R#--- PP:L R#---  A:F n#---  B:F n#--R  A:R
Rw-#-      Rw-#-      w--#-      w--#-      n--#-      n--#-      ---#-      ---#-
-----      -----      -----      -----      -----      -----      -----      -----
```

581
PP:RIGHT I Turned Right Q
A:WAIT Good You Turn Right CQ
582
PP:RIGHT You Move Forward CQ
A:PAT Bad You Move Forward CQ
Invalid command.
583
PP:FORWARD Next Turn Is Yours So You Push Cake CQ
A:PAT Bad You Move Forward CQ
Bad command.
584
PP:FORWARD I Moved Forward Q
A:PAT Good You Move Forward CQ

```
W---- 581  N---- 581  N---- 582  E---- 582  E---- 583  -E--- 583  -E--- 584  --E-- 584
e#--R PP:R e#--R  A:W e#--R PP:R e#--R  A:P e#--R PP:F e#--R  A:P e#--R PP:F e#--R  A:P
---#-      ---#-      ---#-      ---#-      ---#-      ---#-      ---#-      ---#-
-----      -----      -----      -----      -----      -----      -----      -----
```

585
PP:FORWARD I Moved Forward Q
A:PAT Good You Move Forward CQ
586
PP:FORWARD I Moved Forward Q
A:WAIT Good You Turn Right CQ
A is out-of-reach to Pat.
587
PP:RIGHT I Turned Right Q
A:WAIT Good You Push Roll CQ
A is out-of-reach to Pat.
588
PP:FORWARD I Pushed Roll Q
A:WAIT Good You Push Roll CQ
A is out-of-reach to Pat.

```
--E-- 585  ---E- 585  ---E- 586  ----E 586  ----E 587  ----S 587  ----S 588  ----- 588
e#--R PP:F e#--R  A:P e#--R PP:F e#--R  A:W e#--R PP:R e#--R  A:W e#--R PP:F e#--S  A:W
---#-      ---#-      ---#-      ---#-      ---#-      ---#-      ---#-      ---#R
-----      -----      -----      -----      -----      -----      -----      -----
```

589
PP:FORWARD I Pushed Cake Q
A:WAIT Bad You Push Roll CQ
Pushed Roll not Cake.
A is out-of-reach to Pat.
590
PP:FORWARD I Pushed Roll Q
A:RIGHT Good You Turn Right CQ
A is out-of-reach to Pat.
591
PP:RIGHT I Turned Right Q
A:WAIT Good You Push Roll CQ
A is out-of-reach to Pat.
592
PP:FORWARD I Pushed Roll Q
A:WAIT Good Next Turn Is Mine So You Push Roll CQ
A is out-of-reach to Pat.

```
----- 589  ----- 589  ----- 590  ----- 590  ----- 591  ----- 591  ----- 592  ----- 592
e#--S PP:F e#---  A:W e#--- PP:F e#---  A:R s#--- PP:R s#---  A:W s#--- PP:F s#---  A:W
---#R      ---#S      ---#S      ---#-      ---#-      ---#-      ---#-      ---#-
-----      ----R      ----R      ---RS      ---RS      ---RW      ---RW      --RW-
```

593
PP:FORWARD Next Turn Is Yours So You Move Forward CQ
A:FORWARD Good Next Turn Is Mine So You Push Roll CQ
PP reads **WM** properly.
594
PP:FORWARD Next Turn Is Yours So You Squash Cake CQ
A:FORWARD (Squashed) Bad I Squashed Roll So Next Turn Is <Yours Q
595
PP:RIGHT Good You Squashed Cake So Next Turn Is <Mine Q
A:PAT Bad Q
PP turned away from the Wall.
PP sets up **WM** correctly with <Mine.
596
Tchr changes Roll to Cake. <<<<<<<<<<<<<

================================

Remembering Numbers Steps 645-680, 693-716

645
PP:WAIT What Is Your Number Q? A: Now My Number Is Two Q Teacher's mistake.
A:WAIT What Is Your Number Q? PP: Now My Number Is Three Q A: Good
646
PP:WAIT Now My Number Is Three Q Unasked.
A:PAT Bad Now You Wait CQ
647
PP:WAIT What Is Your Number Q? A: Bad Q
A:PAT Now You Wait CQ
648
PP:WAIT Make Your Q
A:PAT Bad Now You Wait CQ

```
----E 645  ----E 645  ----E 646  ----E 646  ----E 647  ----E 647  ----E 648  ----E 648
-#--e PP:W -#--e  A:W -#--e PP:W -#--e  A:P -#--e PP:W -#--e  A:P -#--e PP:W -#--e  A:P
---#-      ---#-      ---#-      ---#-      ---#-      ---#-      ---#-      ---#-
-C---      -C---      -C---      -C---      -C---      -C---      -C---      -C---
```

649
PP:WAIT Q **A shows B with FLT-Quiet how to answer a Question.**
A:WAIT What Is Your Number Quiet Now My Number Is Three Q
650
PP:WAIT You Squashed Cake So Next Turn Is <Mine Q PP sets up **WM** in wrong place.
A:WAIT Bad Now You Wait CQ
651
PP:WAIT I Turned Right Q
A:PAT Bad Now You Wait CQ
652
PP:WAIT Q
A:WAIT Make Your Number One CQ

```
----E 649  ----E 649  ----E 650  ----E 650  ----E 651  ----E 651  ----E 652  ----E 652
-#--e PP:W -#--e  A:W -#--e PP:W -#--e  A:W -#--e PP:W -#--e  A:P -#--e PP:W -#--e  A:W
---#-      ---#-      ---#-      ---#-      ---#-      ---#-      ---#-      ---#-
-C---      -C---      -C---      -C---      -C---      -C---      -C---      -C---
```

653
PP:WAIT Q
A:WAIT My Number Is <One Q **With PP quiet, this is in context of A's previous response.**
654
PP:WAIT Now You Wait Q
A:PAT Bad Now You Wait CQ
655
PP:WAIT Q
A:PAT Now You Wait CQ
656
PP:WAIT Q
A:WAIT Make Your Number Two CQ

```
----E 653  ----E 653  ----E 654  ----E 654  ----E 655  ----E 655  ----E 656  ----E 656
-#--e PP:W -#--e  A:W -#--e PP:W -#--e  A:P -#--e PP:W -#--e  A:P -#--e PP:W -#--e  A:W
---#-      ---#-      ---#-      ---#-      ---#-      ---#-      ---#-      ---#-
-C---      -C---      -C---      -C---      -C---      -C---      -C---      -C---
```

657
PP:WAIT Q
A:WAIT My Number Is <Two Q
658
PP:WAIT Q
A:WAIT What Is Your Number Quiet Now My Number Is Two Q
659
PP:WAIT Q
A:PAT Now You Wait CQ
660
PP:WAIT Q
A:WAIT Make Your Number Three CQ

```
----E 657  ----E 657  ----E 658  ----E 658  ----E 659  ----E 659  ----E 660  ----E 660
-#--e PP:W -#--e  A:W -#--e PP:W -#--e  A:W -#--e PP:W -#--e  A:P -#--e PP:W -#--e  A:W
---#-      ---#-      ---#-      ---#-      ---#-      ---#-      ---#-      ---#-
-C---      -C---      -C---      -C---      -C---      -C---      -C---      -C---
```

661
PP:WAIT Q
A:WAIT My Number Is <Three Q
662
PP:WAIT Q
A:WAIT What Is Your Number Quiet Now My Number Is One Q
663
PP:WAIT Q
A:WAIT What Is Your Number Q? PP: Now My Number Is Three Q A: Good
664
PP:WAIT Q

```
A:WAIT Make Your Number One CQ
----E 661  ----E 661  ----E 662  ----E 662  ----E 663  ----E 663  ----E 664  ----E 664
-#--e PP:W -#--e  A:W -#--e PP:W -#--e  A:W -#--e PP:W -#--e  A:W -#--e PP:W -#--e  A:W
---#-      ---#-      ---#-      ---#-      ---#-      ---#-      ---#-      ---#-
-C---      -C---      -C---      -C---      -C---      -C---      -C---      -C---
665
 PP:WAIT My Number Is <One CQ                                                   CQ wrong.
 A:PAT Good Now You Wait CQ
666
 PP:WAIT Q m
 A:WAIT What Is Your Number Q?        PP: Now My Number Is One^ CQ          A: Bad   CQ wrong.
667
 PP:WAIT My Number Is One Q                               Unasked and Now omitted but no CQ.
 A:PAT Bad Now You Wait CQ
668
 PP:WAIT Q
 A:WAIT Make Your Number Two CQ
----E 665  ----E 665  ----E 666  ----E 666  ----E 667  ----E 667  ----E 668  ----E 668
-#--e PP:W -#--e  A:P -#--e PP:W -#--e  A:W -#--e PP:W -#--e  A:P -#--e PP:W -#--e  A:W
---#-      ---#-      ---#-      ---#-      ---#-      ---#-      ---#-      ---#-
-C---      -C---      -C---      -C---      -C---      -C---      -C---      -C---
669
 PP:WAIT My Number Is <Two^ Q
 A:PAT Good Now You Wait CQ
670
 PP:WAIT Q
 A:WAIT What Is Your Number Q?        PP: Now My Number Is Two^ Q           A: Good
671
 PP:WAIT Q
 A:PAT Now You Wait CQ
672
 PP:WAIT Q
 A:WAIT Make Your Number Three CQ
----E 669  ----E 669  ----E 670  ----E 670  ----E 671  ----E 671  ----E 672  ----E 672
-#--e PP:W -#--e  A:P -#--e PP:W -#--e  A:W -#--e PP:W -#--e  A:P -#--e PP:W -#--e  A:W
---#-      ---#-      ---#-      ---#-      ---#-      ---#-      ---#-      ---#-
-C---      -C---      -C---      -C---      -C---      -C---      -C---      -C---
673
 PP:WAIT My Number Is <Three Q
 A:PAT Good You Turn Left CQ
674
 PP:LEFT I Turned Left Q
 A:PAT Good You Turn Left CQ
675
 PP:LEFT I Turned Left Q
 A:RIGHT Good I Turned Right Q                                                  B saw A turn.
676
 PP:FORWARD I Moved Forward Q                                                 A didn't see it.
 A:FORWARD I Moved Forward Q                                               B didn't see A move.
----E 673  ----E 673  ----E 674  ----N 674  ----N 675  ----W 675  ----W 676  ---W- 676
-#--e PP:W -#--e  A:P -#--e PP:L -#--e  A:P -#--e PP:L -#--e  A:R -#--s PP:F -#--s  A:F
---#-      ---#-      ---#-      ---#-      ---#-      ---#-      ---#-      ---#-
-C---      -C---      -C---      -C---      -C---      -C---      -C---      -C---
...
...
693
 PP:WAIT I Turned Left Q
 A:PAT Bad Now You Wait CQ
694
 PP:WAIT Now Q
 A:PAT Bad Now You Wait CQ
695
 PP:WAIT What Is <Two CQ
 A:PAT Bad Now You Wait CQ
696
 PP:WAIT Make Q
 A:PAT Bad Now You Wait CQ
----- 693  ----- 693  ----- 694  ----- 694  ----- 695  ----- 695  ----- 696  ----- 696
-#--C PP:W -#--C  A:P -#--C PP:W -#--C  A:P -#--C PP:W -#--C  A:P -#--C PP:W -#--C  A:P
W--#-      W--#-      W--#-      W--#-      W--#-      W--#-      W--#-      W--#-
w----      w----      w----      w----      w----      w----      w----      w----
697
 PP:WAIT My Q
 A:PAT Bad Now You Wait CQ
698
 PP:WAIT Q
 A:WAIT Make Your Number Two CQ
```

699
PP:WAIT My Number Is <Three Q   Wrong number.
A:PAT Bad Now You Wait CQ
700
PP:WAIT Q
A:WAIT What Is Your Number Q?   PP: Now My Number Is Three CQ   A: Good
----- 697 ----- 697 ----- 698 ----- 698 ----- 699 ----- 699 ----- 700 ----- 700
-#--C PP:W -#--C A:P -#--C PP:W -#--C **A:W** -#--C PP:W -#--C A:P -#--C PP:W -#--C **A:W**
W--#-   W--#-   W--#-   W--#-   W--#-   W--#-   W--#-   W--#-
w----   w----   w----   **w----**   **w----**   w----   w----   **w----**
701
PP:WAIT Now Q
A:PAT Bad Now You Wait CQ
702
PP:WAIT Q
A:WAIT Make Your Number Two CQ
703
PP:WAIT My Number Is <One CQ   CQ is wrong.
A:PAT Bad Now You Wait CQ
704
PP:WAIT Q
A:WAIT What Is Your Number Q?   PP: Now My Number Is One^ Q   A: Good
----- 701 ----- 701 ----- 702 ----- 702 ----- 703 ----- 703 ----- 704 ----- 704
-#--C PP:W -#--C A:P -#--C PP:W -#--C **A:W** -#--C PP:W -#--C A:P -#--C PP:W -#--C **A:W**
W--#-   W--#-   W--#-   W--#-   W--#-   W--#-   W--#-   W--#-
w----   w----   w----   **w----**   **w----**   w----   w----   **w----**
705
PP:WAIT What Q
A:PAT Bad Now You Wait CQ
706
PP:WAIT Q
A:WAIT Make Your Number Two CQ
707
PP:WAIT My Number Is <Two Q
A:PAT Good Now You Wait CQ
708
PP:WAIT Q
A:WAIT What Is Your Number Q?   PP: Now My Number Is Two^ Q   A: Good
----- 705 ----- 705 ----- 706 ----- 706 ----- 707 ----- 707 ----- 708 ----- 708
-#--C PP:W -#--C A:P -#--C PP:W -#--C **A:W** -#--C PP:W -#--C A:P -#--C PP:W -#--C **A:W**
W--#-   W--#-   W--#-   W--#-   W--#-   W--#-   W--#-   W--#-
w----   w----   w----   **w----**   **w----**   w----   w----   **w----**
709
PP:WAIT What Is <Two Q
A:PAT Bad Now You Wait CQ
710
PP:WAIT Q
A:WAIT Make Your Number One CQ
711
PP:WAIT My Number Is <One Q
A:PAT Good Now You Wait CQ
712
PP:WAIT Q
A:WAIT What Is Your Number Q?   PP: Now My Number Is One Q   A: Good
----- 709 ----- 709 ----- 710 ----- 710 ----- 711 ----- 711 ----- 712 ----- 712
-#--C PP:W -#--C A:P -#--C PP:W -#--C **A:W** -#--C PP:W -#--C A:P -#--C PP:W -#--C **A:W**
W--#-   W--#-   W--#-   W--#-   W--#-   W--#-   W--#-   W--#-
w----   w----   w----   **w----**   **w----**   w----   w----   **w----**
713
PP:WAIT Q
A:PAT Now You Wait CQ
714
PP:WAIT Q
A:WAIT Make Your Number Three CQ
715
PP:WAIT My Number Is <Three Q
A:PAT Good Now You Wait CQ
716
PP:WAIT Q
A:WAIT What Is Your Number Q?   PP: Now My Number Is Three CQ   A: Good
----- 713 ----- 713 ----- 714 ----- 714 ----- 715 ----- 715 ----- 716 ----- 716
-#--C PP:W -#--C A:P -#--C PP:W -#--C **A:W** -#--C PP:W -#--C A:P -#--C PP:W -#--C **A:W**
W--#-   W--#-   W--#-   W--#-   W--#-   W--#-   W--#-   W--#-
w----   w----   w----   **w----**   **w----**   w----   w----   **w----**

=================================

Touching. Steps 945-972, 989-997

945
PP:FORWARD I Moved Forward Q
A:PAT Good You Turn Right CQ
946
PP:RIGHT I Turned Right Q
A:FORWARD Good Now You Wait CQ
947
PP:WAIT Look I Am Touching Black Stump Q
A:RIGHT Good Now You Wait CQ
948
PP:WAIT Look I Am Touching Black Stump Q
A:WAIT What Am I Touching Q? PP: Look I Am Touching Black Stump Q A: Bad

```
C---- 945  C---- 945  C---- 946  C---- 946  C---- 947  C---- 947  C---- 948  C---- 948
-#--S PP:F -#---  A:P -#--- PP:R -#---  A:F -#--- PP:W -#---  A:R -#--- PP:W -#---  A:W
---#-      ---#S      ---#S      ---#W      --n#W      --n#W      --e#W      --e#W
--n--      --n--      --n--      --n--      -----      -----      -----      -----
```

949
PP:LEFT I Turned Left Q
A:WAIT Good You Turn Right CQ
950
PP:RIGHT I Turned Right Q
A:WAIT What Am I Touching Q? PP: Look You Are Touching Black Stump Q A: Good
951
PP:LEFT I Turned Left Q
A:WAIT Good You Turn Right CQ
952
PP:RIGHT I Turned Right Q
A:WAIT What Am I Touching Q? PP: Look You Are Touching Black Stump Q A: Good

```
C---- 949  C---- 949  C---- 950  C---- 950  C---- 951  C---- 951  C---- 952  C---- 952
-#--- PP:L -#---  A:W -#--- PP:R -#---  A:W -#--- PP:L -#---  A:W -#--- PP:R -#---  A:W
--e#W      --e#S      --e#S      --e#W      --e#W      --e#S      --e#S      --e#W
-----      -----      -----      -----      -----      -----      -----      -----
```

953
PP:LEFT I Turned Left Q
A:WAIT Good You Turn Right CQ
954
PP:RIGHT I Turned Right Q
A:PAT Good Now You Wait CQ
955
PP:WAIT Look I Am Touching Black Stump Q
A:WAIT What Are You Touching Q? PP: Look I Am Touching Black Stump Q A: Good
956
PP:LEFT I Turned Left Q
A:WAIT Good You Turn Right CQ

```
C---- 953  C---- 953  C---- 954  C---- 954  C---- 955  C---- 955  C---- 956  C---- 956
-#--- PP:L -#---  A:W -#--- PP:R -#---  A:P -#--- PP:W -#---  A:W -#--- PP:L -#---  A:W
--e#W      --e#S      --e#S      --e#W      --e#W      --e#W      --e#W      --e#S
-----      -----      -----      -----      -----      -----      -----      -----
```

957
PP:RIGHT I Turned Right Q
A:PAT Good Now You Wait CQ
958
PP:WAIT Look I Am Touching Black Stump Q
A:PAT Good Now You Wait CQ
959
PP:WAIT Q
A:RIGHT You Turn Right CQ
960
PP:RIGHT I Moved Forward CQ
A:FORWARD Bad I Moved Forward Q B didn't see A move.

```
C---- 957  C---- 957  C---- 958  C---- 958  C---- 959  C---- 959  C---- 960  C---- 960
-#--- PP:R -#---  A:P -#--- PP:W -#---  A:P -#--- PP:W -#---  A:R -#--- PP:R -#---  A:F
--e#S      --e#W      --e#W      --e#W      --e#W      --e#W      --s#W      --s#N
-----      -----      -----      -----      -----      -----      -----      -----
```

961
PP:FORWARD I Moved Forward CQ CQ is wrong.
A:LEFT Bad You Turn Left CQ
962
PP:LEFT I Turned Left Q
A:FORWARD Good You Move Forward CQ
963
PP:FORWARD I Moved Forward Q
A:LEFT Good You Turn Left CQ

```
964
 PP:LEFT I Turned Left CQ                                                    CQ is wrong.
 A:PAT Bad Now You Wait CQ
C---- 961  C---- 961  C---- 962  C---- 962  C---- 963  C---- 963  C---- 964  C---- 964
-#--- PP:F -#--N  A:L -#--N PP:L -#--W  A:F -#--W PP:F -#-W-  A:L -#-W- PP:L -#-S-  A:P
---#N      ---#-      ---#-      ---#-      ---#-      ---#-      ---#-      ---#-
--s--      --s--      --e--      --e--      ---e-      ---e-      ---n-      ---n-
965
 PP:WAIT Look I Am Touching Black Stump Q
 A:PAT Good Now You Wait CQ
966
 PP:WAIT Look I Am Touching Black Stump Q
 A:WAIT What Am I Touching Q?     PP: Look You Are Touching Black Stump Q     A: Good.
967
 PP:LEFT I Turned Left Q
 A:SLAP Good You Turn Right CQ            A Slaps to stop PP from keeping on turning Left.
968
 PP:RIGHT I Turned Right Q
 A:PAT Good Now You Wait CQ
C---- 965  C---- 965  C---- 966  C---- 966  C---- 967  C---- 967  C---- 968  C---- 968
-#-S- PP:W -#-S-  A:P -#-S- PP:W -#-S-  A:W -#-S- PP:L -#-E-  A:S -#-E- PP:R -#-S-  A:P
---#-      ---#-      ---#-      ---#-      ---#-      ---#-      ---#-      ---#-
---n-      ---n-      ---n-      ---n-      ---n-      ---n-      ---n-      ---n-
969
 PP:WAIT Look I Am Touching Black Stump Q
 A:WAIT Look I Am Touching Black Stump Q
970
 PP:LEFT I Turned Left Q
 A:SLAP Good You Turn Right CQ
971
 PP:RIGHT I Turned Right Q
 A:PAT Good Now You Wait CQ
972
 PP:WAIT Look I Am Touching Black Stump Q
 A:WAIT Look You Are Touching Black Stump Q
C---- 969  C---- 969  C---- 970  C---- 970  C---- 971  C---- 971  C---- 972  C---- 972
-#-S- PP:W -#-S-  A:W -#-S- PP:L -#-E-  A:S -#-E- PP:R -#-S-  A:P -#-S- PP:W -#-S-  A:W
---#-      ---#-      ---#-      ---#-      ---#-      ---#-      ---#-      ---#-
---n-      ---n-      ---n-      ---n-      ---n-      ---n-      ---n-      ---n-
...
...
989
 B:RIGHT I Pushed Cake Q
 A:SLAP Bad You Turn Right CQ
990
 PP:LEFT I Turned Left Q
 A:SLAP Good You Turn Left CQ
991
 PP:FORWARD I Pushed Cake CQ                                                 CQ is wrong.
 A:WAIT Bad You Turn Left CQ                                   A was out-of-reach to Slap.
992
 PP:FORWARD I Pushed Cake Q
 A:LEFT Good You Turn Left CQ                                  A was out-of-reach to Slap.
-s--- 989  -s--- 989  -s--- 990  -s--- 990  -s--- 991  -s--- 991  -s--- 992  -s--- 992
-#---  B:R -#---  A:S -#--- PP:L -#---  A:S -#--- PP:F -#---  A:W -#--- PP:F -#---  A:L
---#-      ---#-      ---#-      ---#-      ---#-      ---#-      ---#-      ---#C
--EC-      --SC-      --SC-      --EC-      --EC-      ---EC      ---EC      ----E
993
 PP:LEFT I Turned Left Q
 A:FORWARD Good You Push Cake CQ                                A is out-of-reach to Pat.
994
 PP:FORWARD Next Turn Is Mine So You Move Forward CQ                          PP uses WM.
 A:FORWARD Good Next Turn Is Yours So You Push Cake CQ
995
 PP:FORWARD Next Turn Is Mine So You Wait Q                                   Q is wrong.
 A:WAIT Good Next Turn Is Yours So You Squash Cake CQ
996
 PP:FORWARD (Squashed) I Squashed Cake So Next Turn Is <Yours Q     PP sets up WM with <Yours.
 A:RIGHT Good You Squashed Cake So Next Turn Is <Mine Q
-e--- 993  -e--- 993  --e-- 994  --e-- 994  ---e- 995  ---eC 995  ---eC 996  -C-eN 996
-#--- PP:L -#---  A:F -#--- PP:F -#--C  A:F -#--C PP:F -#--N  A:W -#--N PP:F -#---  A:R
---#C      ---#C      ---#C      ---#N      ---#N      ---#-      ---#-      ---#-
----E      ----N      ----N      -----      -----      -----      -----      -----
997-11535
 PP:RIGHT I Turned Right Q
                    ============  End of First Interaction ============
```

Second Interaction - No Approval from Teacher

```
129
 PP:RIGHT I m Turned m Right m Q m
 A:FORWARD I m Pushed m Cake o Q m                                         A didn't see it.
130
 PP:FORWARD I m Moved m Forward m Q m
 A:FORWARD I m Pushed m Cake o Q m                                         A didn't see it.
131
 PP:FORWARD I m Moved m Forward m Q m
 A:RIGHT I m Turned m Right m Q m                                          A didn't see it.
132
 PP:FORWARD I m Moved m Forward m Q m
 A:FORWARD I m Pushed m Cake o Q m                                             No approval.
----- 129  ----- 129  ----- 130  ----- 130  ----- 131  ----- 131  ----- 132  ----- 132
-#--- PP:R -#---  A:F -#--- PP:F -#---  A:F -#--- PP:F -#---  A:R -#--- PP:F -#---  A:F
---#-      ---#-      ---#-      ---#-      C--#-      C--#-      C--#-      C--#-
-Cw-S      -Cw-W      Cw--W      Cw-W-      w--W-      w-W--      n-W--      nW---133
 PP:FORWARD I m Moved m Forward m Q m
 A:FORWARD I m Pushed m Cake o Q m                                         A didn't see it.
134
 PP:RIGHT I m Turned m Right m Q m
 A:FORWARD I m Pushed m Cake o Q m                                         A didn't see it.
135
 PP:FORWARD I m Moved m Forward m Q m
 A:RIGHT I m Turned m Right m Q m                                          A didn't see it.
136
 PP:FORWARD I m Moved m Forward m Q m
 A:FORWARD I m Pushed m Cake o Q m                                             No approval.
----- 133  ----- 133  C---- 134  C---- 134  nC--- 135  nC--- 135  eC--- 136  eC--- 136
C#--- PP:F C#---  A:F n#--- PP:R n#---  A:F -#--- PP:F -#---  A:R -#--- PP:F N#---  A:F
n--#-      n--#-      ---#-      ---#-      ---#-      N--#-      N--#-      ---#-
-W---      W----      W----      N----      N----      -----      -----      -----
137
 PP:FORWARD I m Moved m Forward m Q m
 A:FORWARD I m Pushed m Cake o Q m                                         A didn't see it.
138
 PP:RIGHT I m Turned m Right m Q m
 A:FORWARD I m Pushed m Cake o Q m                                         A didn't see it.
139
 PP:FORWARD I m Moved m Forward m Q m
 A:FORWARD I m Pushed m Cake o Q m                                         A didn't see it.
140
 PP:FORWARD I m Moved m Forward m Q m
 A:RIGHT I m Turned m Right m Q m                                          A didn't see it.
-eC-- 137  NeC-- 137  N-eC- 138  E-eC- 138  E--eC 139  -E-eC 139  -E--e 140  --E-e 140
N#--- PP:F -#---  A:F -#--- PP:R -#---  A:F -#--- PP:F -#---  A:F -#--C PP:F -#--C  A:R
---#-      ---#-      ---#-      ---#-      ---#-      ---#-      ---#-      ---#-
-----      -----      -----      -----      -----      -----      -----      -----
141
 PP:FORWARD I m Moved m Forward m Q m
 A:FORWARD I m Pushed m Cake o Q m                                             No approval.
142
 PP:FORWARD I m Moved m Forward m Q m
 A:FORWARD I m Pushed m Cake o Q m                                         A didn't see it.
143
 PP:RIGHT I m Turned m Right m Q m
 A:FORWARD I m Pushed m Cake o Q m                                         A didn't see it.
144
 PP:FORWARD I m Moved m Forward m Q m
 A:RIGHT I m Turned m Right m Q m                                          A didn't see it.
--E-s 141  ---Es 141  ---E- 142  ----E 142  ----E 143  ----S 143  ----S 144  ----- 144
-#--C PP:F -#--C  A:F -#--s PP:F -#--s  A:F -#--- PP:R -#---  A:F -#--- PP:F -#--S  A:R
---#-      ---#-      ---#C      ---#C      ---#s      ---#s      ---#-      ---#-
-----      -----      -----      -----      ----C      ----C      ---Cs      ---Cs
145
 PP:FORWARD I m Moved m Forward m Q m
 A:FORWARD I m Pushed m Cake o Q m                                             No approval.
146
 PP:FORWARD I m Moved m Forward m Q m
 A:FORWARD I m Pushed m Cake o Q m                                         A didn't see it.
PLAN: 1-1  PP:RIGHT I m Turned m Right m Q m
      1-6  A:FORWARD I m Pushed m Cake o Q m
      2-11 PP:FORWARD                         End of plan with novelty goal.  11 substeps.
147
 PP:RIGHT I m Turned m Right m Q m
 A:FORWARD I m Pushed m Cake o Q m                                         A didn't see it.
```

```
148
 PP:FORWARD I m Moved m Forward m Q m
 A:FORWARD I m Pushed m Cake o Q m             Plan followed for 11 substeps. A didn't see it.
----- 145  ----- 145  ----- 146  ----- 146  ----- 147  ----- 147  ----- 148  ----- 148
-#--S PP:F -#---  A:F -#--- PP:F -#---  A:F -#--- PP:R -#---  A:F -#--- PP:F -#---  A:F
---#-      ---#S      ---#S      ---#-      ---#-      ---#-      ---#-      ---#-
---Cw      ---Cw      --Cw-      --CwS      -Cw-S      -Cw-W      Cw--W      Cw-W-
149
 PP:FORWARD I m Moved m Forward m Q m                                          A didn't see it.
 A:RIGHT I m Turned m Right m Q m
150
 PP:FORWARD I m Moved m Forward m Q m
 A:FORWARD I m Pushed m Cake o Q m
151
 PP:FORWARD I m Moved m Forward m Q m                                          A didn't see it.
 A:FORWARD I m Pushed m Cake o Q m
PLAN: 1-1  PP:RIGHT I m Turned m Right m Q m
      1-6  A:FORWARD I m Pushed m Cake o Q m
      2-11 PP:FORWARD I m Moved m Forward m Q m
      2-16 A:RIGHT I m Turned m Right m Q m
      3-21 PP:FORWARD I m Moved m Forward m Q m
      3-26 A:FORWARD I m Pushed m Cake o Q m
      4-31 PP:FORWARD I m Moved m Forward m Q m
      4-36 A:FORWARD I m Pushed m Cake o Q m
      5-41 PP:RIGHT I m Turned m Right m Q m
      5-46 A:FORWARD I m Pushed m Cake o Q m
      6-51 PP:FORWARD I m Moved m Forward m Q m
      6-56 A:FORWARD I m Pushed m Cake o Q m
      7-61 PP:FORWARD I m Moved m Forward m Q m
      7-66 A:RIGHT I m Turned m Right m Q m
      8-71 PP:FORWARD I m Moved m Forward m Q m
      8-76 A:FORWARD I m Pushed m Cake o Q m
      9-81 PP:FORWARD I m Moved m Forward m Q m
      9-86 A:FORWARD I m Moved m Forward m Q        Error !!!
     10-91 PP:RIGHT                                End of plan with novelty goal.  91 substeps.
152
 PP:RIGHT I m Turned m Right m Q m                                             A didn't see it.
 A:FORWARD I m Pushed m Cake o Q m
----- 149  ----- 149  ----- 150  ----- 150  ----- 151  ----- 151  C---- 152  C---- 152
-#--- PP:F -#---  A:R -#--- PP:F -#---  A:F C#--- PP:F C#---  A:F n#--- PP:R n#---  A:F
C--#-      C--#-      C--#-      C--#-      n--#-      n--#-      ---#-      ---#-
w--W-      w-W--      n-W--      nW---      -W---      W----      W----      N----
153
 PP:FORWARD I m Moved m Forward m Q m                                          A didn't see it.
 A:RIGHT I m Turned m Right m Q m
154
 PP:FORWARD I m Moved m Forward m Q m
 A:FORWARD I m Pushed m Cake o Q m
155
 PP:FORWARD I m Moved m Forward m Q m                                          A didn't see it.
 A:FORWARD I m Pushed m Cake o Q m
156
 PP:RIGHT I m Turned m Right m Q m                                             A didn't see it.
 A:FORWARD I m Pushed m Cake o Q m
nC--- 153  nC--- 153  eC--- 154  eC--- 154  -eC-- 155  NeC-- 155  N-eC- 156  E-eC- 156
-#--- PP:F -#---  A:R -#--- PP:F N#---  A:F N#--- PP:F -#---  A:F -#--- PP:R -#---  A:F
---#-      N--#-      N--#-      ---#-      ---#-      ---#-      ---#-      ---#-
N----      -----      -----      -----      -----      -----      -----      -----
157
 PP:FORWARD I m Moved m Forward m Q m                                          A didn't see it.
 A:FORWARD I m Pushed m Cake o Q m
158
 PP:FORWARD I m Moved m Forward m Q m                                          A didn't see it.
 A:RIGHT I m Turned m Right m Q m
159
 PP:FORWARD I m Moved m Forward m Q m
 A:FORWARD I m Pushed m Cake o Q m
160
 PP:FORWARD I m Moved m Forward m Q m
 A:FORWARD I m Pushed m Cake o Q m   Plan-following ended on Step 10-88:Pushed/Moved mismatch.
E--eC 157  -E-eC 157  -E--e 158  --E-e 158  --E-s 159  ---Es 159  ---E- 160  ----E 160
-#--- PP:F -#---  A:F -#--C PP:F -#--C  A:R -#--C PP:F -#--C  A:F -#--s PP:F -#--s  A:F
---#-      ---#-      ---#-      ---#-      ---#-      ---#-      ---#C      ---#C
-----      -----      -----      -----      -----      -----      -----      -----
                          ==========  End of Second Interaction  ==========
```

Printed in the United States
By Bookmasters